1968

"A Vietnam Love Story"

James Marquis

1968
"A VIETNAM LOVE STORY"

James Marquis

"THE FIRST GAY AMERICAN SOLDIER"

Baron Von Steuben
Continental Army
1778 To 1783

*B*aron von Steuben was a former Persian military officer; he served as Inspector General and Major General of the Continental Army (A Two Star General) during the American Revolution. He was one of the fathers of the Continental Army in teaching them the essentials of military drilling, tactics, and discipline. He wrote the regulation for the order of discipline of the troops of the United States, and the book served as the army's drill manual for decades. He also served as George Washington's Chief of Staff in the final years of the war. He resigned from the service and settled with his longtime companions, William North and Benjamin Walker, until his death on November 28,

1794, at his estate in Oneida County, New York. The estate became part of the town of Steuben, New York, which is named after him.

Von Steuben arrived in the United States with his seventeen-year-old secretary, Peter Stephen Du Ponceau. At Valley Forge, he began a close relationship with Benjamin Walker and William North, then both military officers in their 20s. Some historians believe these extraordinary intense emotional relationships were romantic. Von Steuben treated them as surrogate sons. A third young man by the name of John W. Mulligan also considered himself one of Steuben's sons. He inherited Steuben's vast library, collection of maps and $2,500 in cash. Upon Steuben's death, his remaining property was divided between his military companions North and Walker.

- Wikipedia

CONTENTS

1968

A VIETNAM WAR LOVE STORY

BEFORE

After Jim walked into his apartment after a day's work at the bank, his phone was ringing off the hook. When he answered it, it was his mother. Strangely enough, she didn't ask how his day was or how he was feeling. All she said, with panic in her voice, was, "You got it! Jim said, "Got what?" His mother said, "The letter." Jim said, "What letter?"

"The letter from the selective service board," his mother said, trying to conceal the fear in her voice. Small beads of sweat now appeared on Jim's forehead, as he knew what it was going to say. He told his mother to go ahead and open it and read it to him.

The instructions contained within the letter were, "You are to report for your pre-induction physical at the Chicago induction Center on July 28, 1967. Contact the Army recruiting officer located in Watseka, Illinois, for transportation arrangements and further details."

It was now July 1st, only five days from Jim's 19th birthday. Jim told his mother that he would follow up with the instructions and be in touch with her shortly.

Jim walked to his mini-bar in his apartment and made a stiff Scotch and soda, and sat down to digest all of the information that he had just received during the phone call with his mother. He knew that many of his friends had been drafted during the escalation of the Vietnam crisis. It was just a matter of time before he was called to active duty. Since graduating from high school, he had secured a great position as a teller in a bank in Champaign, which he thoroughly enjoyed.

For over a year, Jim had established himself as the most sought-after teller at the bank because of his glowing personality, his looks and outstanding communication skills. He had a unique talent with people that set him apart from his co-workers as customers would stand in his line, just to spend a few moments with him. They often gave praise to him and would tell management that Jim always brought a smile to their day. His co-workers, many twice his age, were envious of the "PLC" (Personality, Looks and Communication) skills that he possessed.

July 28th was a day that he would never forget. Along with ten other individuals, Jim met at the train station in Watseka to board the three-hour train trip to Chicago. Once they arrived, transportation had already been arranged to transfer them to the massive induction center in downtown Chicago. Once there, they were joined by approximately 300 additional men ready for their pre-induction procedure. They were all herded to various examination areas, usually dressed only in their underwear, socks and shoes. Jim was amazed as to the physique of many of the men, some of which were absolutely beautiful.

At the end of the day, Jim returned to Watseka, got into his car, and drove back to his apartment in Champaign. That night he visualized the whole induction process and the men who surrounded him. He had never been surrounded by so many handsome men. It stimulated both his male hormones and his curiosity. Jim's interest in men had always been there as far back as he could remember, but he had never had the chance to be around so many attractive faces at one time. He was beginning to like the military.

During his last year at the bank, he had been invited to several private parties held at the homes of his male clients. During these gatherings of ten to fifteen people, he would be introduced to the inner circle of the gay population and met with many highly regarded civic leaders in the community. At these cocktail parties, Jim was impressed with his ease in adapting to his new circle of prominent businessmen, whom he met in such a short time.

In two weeks, almost to the day, he received his notice to report for active duty. The report date was stamped for August 28, 1967, only thirty days after him first getting 'the call.' He was amazed at how quickly it came after his pre-induction physical. He realized that the army was deeply engaged in Vietnam, and as a result, the army was drafting massive numbers of men to meet their needs.

When he announced his departure date to the bank, many of his customers planned goodbye parties. Tommy, the bank president, hosted a party at his private country club where all the bank employees attended to share a temporary goodbye to Jim. John, the city's only gay councilman, hosted a huge party for Jim; in attendance were representatives from all the major stores and shops within the gay community.

One of his favorite customers was a reporter for the major newspaper in Champaign. He published a full column article entitled: "The First Union Bank is losing their teller with the biggest smile to the Army on August 28th." Jim was so surprised as to the accolades that so many people bestowed upon him during this time; to him, he was just doing what came naturally. Even his father, who he didn't have much regard for, sat him down and said he would rather go in his place if he could and that Jim should never volunteer for anything while he was in the army. Well, he didn't listen to his father and let the information go in one ear and out the other, as Jim was his own man and could determine what was right for him at any given time.

BASIC TRAINING

*W*hen August 28th arrived, he set off to the Chicago airport, where he boarded a plane and flew to El Paso, Texas, with instructions to report for basic training at Fort Bliss. Upon his arrival, he met with over 300 fellow inductees and were all transported next to the Induction Center. They were divided into formations of ninety inductees by an asshole drill sergeant named 'David Withers,' who had absolutely no respect for any of them. The first question out of the Sergeant Wither's mouth was, do any of you girls want to take a typing test? Immediately, Jim raised his hand, disregarding his father's advice not to volunteer for anything. Rule one of one, already broken.

Jim had taken typing all throughout high school, and he excelled at it. With his ability, he could easily exceed typing speeds of seventy words per minute. The test was administered on one of the worst typewriters that he had ever used in his life! When he walked into the room to take the typing test, there were about twenty manual typewriters scattered around the room. Many of them had keys missing, and Jim had not used a manual typewriter for years. They were asked to leave as soon as they were finished, and the results were never revealed. Now back in formation, they were loaded onto buses and taken to some wooden barracks located in the basic training center of the base. These primitive barracks were not air-conditioned, with rows and rows of bunk beds with no privacy, twenty toilets without dividers, and as many showers in one big room for the recruits to use at any time.

Jim was amazed as the collection of the inductees. All ethnicities, educational backgrounds (no college boys, of course... they were exempt), farmers, storekeepers, guys who just graduated from high school and one banker. Amazingly he got along with all of them because he had a special spark that put people at ease. He liked being around men, and I guess they knew it. To his surprise, he watched these so-called "straight" men checking out other men in the shower. Some of these "macho men" taunted those soldiers who were obviously less masculine. He soon realized that the bully's biggest problem was *fear*.

They were uncomfortable with their surroundings, and as a show of their physical strength, they would pick on individuals not deemed their equals. Jim ended up befriending the bullies, who never picked on him. He told the bullies that within a matter of months, their lives might very well depend upon these individuals that they are picking on and suggested to them that they should stop it. Jim also made friends with many of the black men in the platoon. Up until then, he had never had the opportunity to talk directly to a black person or even seen one naked. He was astounded as to their beautiful bodies, which were so smooth, and generally hung like a stallion. It may have taken a little while, but Jim was playing cards with them and spent time talking about their hardships before their life in the army.

During basic training, you must understand that Jim had never lifted anything heavier than a bundle of $20-dollar bills for over a year before the army and his meager physique was the worst in the entire unit. However, everyone in the platoon saw to it that he passed the physical demands of training, the marching, and the final PT test so that he could graduate on time with the unit. During the 10-week training, Jim became friends with two guys that he assumed were gay, and he ensured that they were not picked on or mistreated by anyone in the platoon. At the end of their training, the unit was a united team, and each member of the unit was ready for reassignment within the regular army.

To Jim's surprise, he was promoted one grade to PFC (Private 1st Class-E2) when he finished basic training. Only eight out of the ninety trainees were promoted. To this day, Jim cannot fully understand why he was promoted over more qualified individuals in the unit. When he thought about it, he reflected back on private discussions with the Company Commander, who praised him on his leadership ability in bringing people together and solving personnel issues within the unit. Maybe this was the reason for his promotion; he didn't know for sure. Liking men was always on his mind, and with this particular gift, he was in his comfort zone within the unit. His comfort was shared by many of the men within the company who would otherwise be troublemakers.

On graduation day, when individual orders were handed out, Jim was assigned to HQ Fort Bliss.

He said to himself, *"Oh shit.... It's unbelievable; this 19-year-old gay guy just hit the jackpot."* Without any advanced training or schooling, he was assigned to personnel, or as those in the military call it, S1, for the entire Army base at Fort Bliss, Texas. Soon after he received these orders, a Jeep pulled up to deliver him to his new assignment.

The Jeep stopped in front of the building called "Command Headquarters." It consisted of four air-conditioned floors containing enlisted records, officer records, finances and the Commanding General's staff. Jim reported to the noncommissioned officer in charge of operations. As he toured the facility, the first thing that came to mind was that it looked like the newsroom of the New York Times. He was so impressed that everyone was working so hard and that they appeared to be enjoying their job. He was turned over to an SP5 (Army rank equal to a sergeant) by the name of Mike, that showed him to his quarters and continued the tour of the operations center.

Jim started working in the Enlisted Records section. Mike saw to it that he received training by the most senior soldier in that section whose name was Andrew. Andrew, slim but athletically built with dusty blonde hair, worked alongside this big burly looking man named Frank. They both worked in the same section and have been side by side for the last eight months. They were extremely knowledgeable and overly friendly. They saw to it that Jim was trained by the best in each section within headquarters administration.

It wasn't long before they asked Jim to join them for one of their evening outings into El Paso. Not to Jim's surprise, the first bar they went to was called *Faces*, which was the most popular gay bar for soldiers stationed at the base. The three of them ended up frequenting that bar several times a week. Andrew and Frank introduce Jim to a few other gay soldiers that worked in headquarters. One of the individuals that Jim met at the bar was Will, who was a distinguished man known by many of the customers at the bar. Andrew, Frank and Jim also recognized Will as a Captain that worked as the aid to the Commanding General of Fort bliss. They knew it was against Army regulations for enlisted personnel to fraternize with officers in the army. But rules were frequently broken.

Will added to the number of clerks that Jim knew, and the advantage of knowing several gay clerks in the office was that they ensured Jim received the absolute best training available while they were there. During the first ninety days, Jim also memorized all of the Army regulations. After four months on the job, he was assigned as the utility clerk to cover all positions in headquarters personnel. After that, Jim was promoted from an E2 or Private First Class, and then again in six months, he rose two ranks to an E4 or Corporal, which was record time during that period. The speed in which he was promoted raised eyebrows among his peers but not his gay friends.

During the same period, Jim's relationship with Will had developed into a casual sexual relationship. He would frequently spend the weekends with him to avoid living in the barracks with his co-workers. Will had a very close relationship with the Commanding General, serving as his administrative assistant, and would frequently join the Commanding General at his home for dinner and drinks.

One weekend Will invited Jim to join him to go to a dinner at the General's house. Jim had always used his personality, looks, and communication skills (PLC) to tackle any situation that he came across in the past. He felt that joining Will for dinner at the General's house would be another slam-dunk situation for him to prove himself. Will introduced him as his cousin because officers could not fraternize with the enlisted soldiers, so he made up the story to cover both their asses. During dinner, a rapid stream of small talk flowed around the table, and the General attempted to zero in on how they were related. Will and Jim felt they satisfactorily answered all of the questions posed to them and left the dinner quite happy.

NAM

$\mathcal{O}$ne downside to being assigned stateside during the Vietnam War was usually during your 11th month in the service. The 11th month was dreaded by many of the troops because that was normally when you would receive your deployment orders to Vietnam. Many of Jim's friends, including Andrew and Frank, had already been reassigned and had left for Vietnam. It seemed like every week, there was another going-away party for more of his friends who were going to Vietnam.

When September finally arrived, which was Jim's 13th month of assignment stateside, he felt relieved. He believed that he would probably not be sent to Vietnam due to an Army policy that barred sending soldiers with eleven months or less left in service to Vietnam. Jim just made the cut and had only eleven months remaining on his tour of duty.

In mid-September, Jim was shocked when he received orders, with just his name on it, to report to Vietnam on September 28, 1968. He discussed this with his co-workers, and they determined that he must have pissed off someone up the chain of command who had enough connections to have an order issued in his name only to report to Vietnam. Will, had already been discharged two months ago. Jim wondered about his dinner with Will and the General, and the General's probing questions about their relationship. All of Jim's friends agreed that the General, who will remain nameless, was the originator of the order for Jim to report for reassignment to Vietnam.

There was nothing that Jim could do at this point but to obey the order. He flew to Fort Lewis Washington, which was the stateside staging center for soldiers in transit to Vietnam. He, along with approximately 250 other soldiers, boarded a Pan American Airlines plane that subsequently landed in Cam Ranh Bay, Vietnam. When the plane landed, Jim stepped off to a sweltering and extremely humid day, one which Jim was anything but accustomed to. He stood sweltering in a long line at a reassignment tent, waiting for his orders for his new duty post.

Sweat poured off his frame in gallons hitting the earth, and with each drop, he felt closer to death. Jim thought to himself, "I'll probably go down in history as the quickest death in Vietnam due to this heat."

When Jim got to within eyeshot of the reassignment desk, he was surprised to see Andrew and Frank, his gay friends from Fort Bliss, handing out orders to the new arrivals. When he got close enough, he yelled out their names, and when they recognized him, Jim thought they were going to jump over the desk and kiss him. But instead, Andrew called out Jim's name and asked him to join them behind the desk. "Jeez, Jim! You look close to death!" Andrew said.

Jim replied, "Yeah, that's because I haven't had my morning scotch yet," and laughed. Frank handed Jim a flask of water, and Jim guzzled it down and poured the last bit over his head. Jim looked exhaustingly over at Frank, and before he could even mutter a word, Frank said, "You'll get used to it."

After Andrew and Frank finished their work that night, they had Jim join them in their hooch, where he stayed for a couple of nights while they searched various units in Vietnam for a "safe" position for his next duty station. Andrew and Frank informed Jim that as soon as they had arrived in Vietnam, that they had become lovers. Jim was very happy for them and told them that he knew that they would eventually become a couple by the way they used to look at each other back on the base.

Jim was eventually assigned to the Headquarters Battery of the 6th/56th Artillery (HHB 6th/56th Arty) located in Chu Lai, just south of Danang. Andrew called ahead to inform them of Jim's arrival date and time, and for them to meet him at the airport. Additionally, they told the Adjutant of the battalion that they had trained Jim at Fort bliss, and he was an exceptionally brilliant soldier in personnel operations. In the army, like any other corporation, the network that surrounds you is invaluable. Having a close relationship with Andrew and Frank at Fort Bliss really paid off when it came time to secure the best possible position in Vietnam.

When Jim arrived in Chu Lai, a Jeep was waiting to pick him up and to deliver him to the battalion personnel office. When Jim walked into the office, the very first person he noticed was the most handsome man that he had ever seen in his entire life. This Greek God got up from his desk and approached him with hand extended and said, "You must be

Jim; they call me **Tito**." Jim looked around the large office, and very good-looking men were everywhere around. Next, Tito introduced him to the Adjutant who got up from his desk and shook Jim's hand and said, "Welcome aboard, we have heard a lot of good things about you."

Tito took Jim to their hooch and instructed him to unpack and get some rest. He said that he would meet Jim for dinner after work, then they should shower and meet with the four other men who live in the hooch so they all could get acquainted with each other. After dinner, Tito and Jim headed to the showers, which Jim desperately needed for he had traveled for almost ten hours that day in the heat and humidity of Vietnam and the smell he produced was legendary.

The hooch was the sleeping quarters for six soldiers. Its outer walls were half wood and the other half screens, and the roof was corrugated metal. Outside of the wood walls were sandbags stacked up to the screened area. There were wooden flaps extended by sticks over the screens propped up, to provide ventilation. The floors of the hooch were plywood, and it was sectioned off with plywood dividers four-feet-high to give each soldier some degree of privacy.

Tito lived in the southeast corner of the hooch next to David and in the same corner where Jim bunked. On the other side, across from Tito, was Ricky, next to him Matthew then Andre. All six of the soldiers were extremely handsome and had fun-loving personalities. Jim fit in instantly. Behind the hooch was a large, man-made bunker used during frequent rocket and mortar attacks, mostly during the night. Also, behind the hooch was the outdoor theater, with benches, where movies were shown once a week.

The first evening, when Jim was visiting with his mates, they told him that he replaced the clerk that had drowned in the ocean while swimming at their beach. This unit was used as the in-country R & R (rest and recuperation) center for our unit. It was located on the Red China Sea, which had crystal blue waters and white sandy beaches. It also contained a lavish nightclub where all the enlisted men were welcome to attend. David, one of our hooch mates, frequently performed for us at our hooch or at the club. He was an extremely talented individual and would play his guitar and sing like Elvis Presley every chance he got. He had the sweetest personality along with handsome good looks, which helped make even the most horrid of nights pass with ease.

From the first day that Jim arrived at headquarters battery, he and Tito were joined at the hip. They work together, took their meals together, and of course, showered together. Tito was 5'11" with a slender but athletically toned body. His face was framed with black hair, brown eyes and a neatly trimmed black mustache. This Puerto Rican from Miami also boasted a 7" uncircumcised piece of manhood hanging proudly between his muscled legs. He was a perfect silhouette of a stately man. His personal grooming and walk were that of a very sophisticated gentleman.

Almost every night, all of the hooch mates would get together and go to the club, drink a few beers, and be entertained by David's unique vocalizations of Elvis Presley songs. Two of Jim's hooch mates, Matthew and Andre, had been assigned to headquarters for three months, and it seemed to Jim that they were in a relationship because, during Jim's first four nights, they would disappear somewhere on the compound and return very early the next morning.

Jim's duties in the office were very similar to those he performed stateside. From the beginning, he was the utility clerk to cover any or all positions that were empty or behind schedule. The Adjutant was duly impressed with Jim's flexibility and knowledge of Army regulations and the speed with which he performed his duties. Tito's duties, on the other hand, were officer records, and he was cross-trained in finance, which made him an invaluable asset.

One night after work, when Tito and Jim were returning to their hooch after their shower, Jim told Tito that he had never seen such a beautiful body in his entire life. Tito told Jim that he was not so bad himself. That was the spark that ignited their relationship from that day forward. That night, after all six of his hooch mates went to bed, Jim was awakened by a set of warm hands covering his mouth and Tito whispering in his ear, "Be quiet."

Tito proceeded to engulf Jim's body with kisses from his lips, to his neck, to his chest, and with his body lying on top of Jim, he smothered him with love. Jim was so surprised; his only reaction was to lay there and open himself up like a book, and to allow Tito to take him like a woman. In the early hours of the morning, Tito returned to his private section of the hooch, and Jim relished in the newfound love that he had just discovered. Never in his wildest dreams did Jim envision this delightful, loving encounter with Tito so early on after arriving at HQ.

The next day at work, the Adjutant called Jim into his office and requested him to visit all of the line units in the battalion to meet their battery clerks and review their personnel procedures. The 6/56th Artillery had five-line units scattered all over Vietnam. From Tan Nh So Nhut, in the South, to Danang, in the North. The first thing Jim thought about was that these trips would cause Tito and him to be separated. Jim contacted the transportation unit, which arranged for brief overnight visits during the next month to all the battalion's units in Vietnam to ensure that he could wrap up his duties as quickly as he can.

The most interesting unit that Jim visited was Delta Battery, located on an island outside of Chu Lia. After taking a twenty-minute boat ride to the island, Jim would travel by a Jeep convoy through several Vietnamese villages to get to the island's mountaintop unit. The battery clerk there was a man named Ron. The only words to describe him were A-list handsome with one of the greatest personalities Jim has ever encountered. Ron stood about an inch shorter than Jim and was of German descent. He was well-built, reminiscent of the statues of Rome, and in Jim's eyes, he was a perfect specimen of a man. Jim said to himself if he were not already involved with

Tito, he may be interested in pursuing a relationship with Ron. But that was a passing thought.

Jim had spoken to Ron many times on the phone, and he knew him as a funny person, plus he was good at his job. After work that night, Ron and Jim went to the club to have a few beers. Ron introduced him to the unit's telephone operator, who they called "Mouse." This guy was short and skinny, but they didn't call him Mouse because of his miniature figure but because of where he worked. Mouse worked in an underground bunker, twelve hours a day and seven days a week as the battalion's telephone operator. It was one hell of a position that not many men could do, but Mouse served his position well and with pride. On that particular night, Ron, Mouse, and Jim proceeded to have several beers, and Mouse got unconditionally plastered. Because Jim was staying overnight, all three of them headed back to Ron's hooch, where he lived alone, and Mouse stripped off all of his clothes and entertained them with lap dances. It was so funny because Mouse was utterly homely. He stood about 5'1", skinny, with a shaved smooth body, with a penis like the tail of a mouse and his face long and narrow with a peaked nose... so okay, he looked like a mouse as well, but he was absolutely hilarious as he bounced around naked.

Later that night, after Mouse had departed, Ron made up a cot next to him where Jim was going to sleep. Both of them had drunk a lot of alcohol and were anything but bashful when they stripped down to their underwear before going to bed. Jim found it interesting that Ron was wearing bikini underpants that emphasized a huge bulge in the front and a bubble ass. Ron was not bashful in the least and showing off his stunning physique in an attempt to gain Jim's attention. Jim thought for a minute on how to diffuse this sexual situation that he was in because he was and, at this point in time, head over heels in love with Tito.

Jim just came right out and said: "Ron, you are really handsome. I bet you get lonely out here on this island. But I must admit if I were not already in a relationship, I would definitely enjoy exploring your "manly parts." Ron just snapped his fingers and said, "Darn it, at least I tried." They both laughed and crawled under the covers and slept the night away. That morning when they both awoke, Ron had a morning boner that was so huge that it came out of his bikini underwear. Jim couldn't help but notice it and told Ron to tuck it in because it was causing Jim's sexual frustration.

After Jim completed the review of Delta battery's personnel procedures, he made the return trip back to HHQ and was happily reunited with Tito that night. Jim found these visits to line units a personal distraction to his relationship with Tito. The only saving grace was when they got together on the eve of his return. Their passionate lovemaking was so intense it seemed like Jim had been gone for months instead of a day or two. Jim told Tito about the encounter with Ron at Delta Battery and that he had a manly physique and how he did everything within his power to seduce him while he was there. But Jim assured Tito that all attempts that Ron made failed.

It was about fourteen days after his return from Delta battery that Jim received notification that a court-martial was going to be convened for one of the enlisted men on the island. Mickey Jones was charged with the attempted murder of the First Sergeant. After a detailed phone conversation with Ron, he discovered that Mickey was, in fact, Mouse -- that skinny guy who had entertained them while Jim was on the island. Mickey's dislike for the First Sergeant was widely known and understood on the island. The First Sergeant had caught Mouse on several occasions performing lap dances in a naked state on various soldiers, and he despised Mouse's sexual displays of nudity in his unit. He had confined Mouse to the island on numerous occasions as punishment for his outlandish behavior.

The court-martial documents revealed that Mickey had wired a hand grenade to the First Sergeant's hooch door, hoping that it would go off and kill him when he entered. Apparently, the hand grenade was a dud, and the investigation by the military police identified Mickey as the culprit. When the court-martial was convened, Mickey was found guilty of attempted murder and sentenced to ten years of hard labor at Fort Leavenworth, Kansas. This was only the second court-martial that Jim had handled during his time in the army. But he was the only clerk in personnel headquarters that was trained in the art of handling military punishment such as article 15's and courts-martial. The Adjutant was very impressed with Jim's ability to handle such complicated legal proceedings and recognized Jim as an invaluable part of his team personally.

It was now two months since Jim has been assigned to HHQ. He and Tito's relationship was in full swing, and it had become obvious to their co-workers and many in the battalion, including the Adjutant, that they were more than friends. Every night, Tito and Jim, along with their hooch mates, would go to the club and be entertained by David singing his Elvis Presley songs. When Tito and Jim returned to the hooch, oftentimes, they would retreat to the bunker for their night of intense lovemaking that at times lasted until the morning. To Jim, Tito was the very embodiment of perfection. An individual whose grace and refinement enhanced whatever he touched, including Jim himself, for he was his master craftsmen. Slowly but surely etching Jim with the knowledge, the confidence and the love that Jim needed to succeed in any climate.

It was also during these retreats that Tito would make plans for them to share their life together after Vietnam. His father owned a very large import/export business in Miami, Florida, at which he said that they would work together after their tours were completed. Jim was most agreeable and looking forward to spending his life with Tito and his family. Nothing could have been better than spending his entire life with this man.

When Christmas of 1968 rolled around, it was the best Christmas that Jim had ever celebrated. One of the overweight MPs (military police) dressed as Santa Claus and passed gifts of booze, treats from back home and magazines. Some of the guys at the club would sit on his lap and tell them what they wanted for Christmas. It was so funny when Jim got on Santa's lap; he told him that all he wanted for Christmas was to be with Tito for the rest of his life. When Santa Claus repeated his wish, the entire club roared of laughter, and Jim and Tito received a standing ovation. That Christmas Eve, Tito and Jim were joined by Matthew, Andre, Ricky and David in the bunker for a night of monogamous but shared lovemaking that never will be forgotten. This was the first time that all three couples had sex in front of their comrades, and it was quite entertaining, to say the least. All three couples played a little game to see which one could make it last the longest, and of course, Jim and Tito won.

New Year's Eve, 1968 -- what a night! The gang of six headed back towards Jim's hooch, with drinks in hand, while welcoming 1969 in with a medley of songs led by David and his guitar. Staggering as they walked, they had just made it back to their hooch when all hell broke loose. A barrage of rockets and mortars began landing in various areas within the compound. All six of them very much intoxicated, ran as best they could to their bunker. Once inside, they could hear a crashing blow that destroyed the screen of their outdoor movie theater located only a few feet from where they laid. Once all the explosions had stopped, all of them instantly sobered up with the realization of how close they had come to losing their lives that New Year's morning. This was still a warzone. After everything had quieted down, they proceeded to give thanks to their lovers in the most gratifying manner possible, making love as if it would be there last time together.

That New Year's Day, the same six guys decided they would head to the beach for the day. All of them took their poncho liners so they could use them as blankets as they proceeded to the south end of the beach, which was designated as clothing-optional. What a beautiful sight it was for them all, for the manhood of all of the soldiers romping in and out of the surf and at times partially erect by just the sight of their mates surrounding them. When they returned to the hooch that night, all except Tito was sunburnt. Tito's bronze skin was only deliciously darkened by the rays of the South Vietnamese sun.

They were all excited to have New Year's Day off from work. That was the first time since Jim had been there that they all were able to enjoy each other's company for an entire day. They all agreed that 1969 was going to be a great year because all of them would be able to finish their tour in Vietnam. Jim and Tito were probably the happiest of the group because every day, they were making plans for their life together.

RIBBONS

*O*n January 9, 1969, Jim was called into the Adjutant's office. Jim was informed that his grandfather had passed away. He was named after his grandfather, who was a very successful businessman despite having only a second-grade education. His grandfather was also very homophobic. While Jim was growing up, his grandfather would tease him about being somewhat feminine and that he would castrate him just like he does his pigs if he didn't straighten up. So, at an early age, Jim had to learn how to adapt to a homophobic environment and live around a successful businessman. Jim never resented his grandfather for this attitude; instead, he was grateful that he taught him how to strengthen his character, steel himself to any insults, and live his life as he does today.

During January 1969, the noncommissioned officer in charge of personnel had a nervous breakdown, and his Sergeant First Class (E7) position was left unfilled. The Adjutant called Jim to his office and asked him if he would like to fill the vacant position. He also stated that if he took the position, he would be promoted to a sergeant immediately and would have total operating responsibility for personnel. As Jim was trained by the best at Fort Bliss, he was more than willing to take the position and was promoted to sergeant with little more than fifteen months of active duty. His co-workers were delighted to have him in charge, and now Tito reported to him.

Headquarters was located on a hill above Chu Lai airport, which housed numerous fighter jets and transport airplanes. The airport was the target for numerous rocket and mortar attacks, which were generally at night. The rockets and mortars were intended for the airport but occasionally their unit instead. When the attacks started, they would all head for the bunkers for safety as they did on that New Year's Eve. Unfortunately, these attacks were so late at night, and they began to take them for granted. Depending on the number of drinks they had on any given night, their response time to get in the bunker was not always too quick. After a while, rocket and mortar attacks became part of their daily routine.

Now, as the noncommissioned officer in charge of personnel, Jim interacted daily with brigade headquarters in Chu Lia. He developed personal working relationships with many of the high-ranking officers in command. His communication skills enabled him to network through various division offices to obtain additional allocations for R&R's along with promotion allocations and other needed supplies for his battalion. He became well known by battalion officers who would call Jim to arrange for various favors that they could not get through any other channels, even their own.

Tito was amazed by Jim's networking ability. He would often comment that his skills would be a real compliment to his father's import/export business. At night, before or after their periods of intense lovemaking, Tito would discuss their future together. One night, Tito told Jim that he never had, in all of his life, had a loving relationship with another man. He had spent the majority of his high school years playing tennis and was recognized as the Florida State Champion. As a result, he spent all of his free time on the tennis court. Jim was bewildered by this. Where and how did Tito learn to express love so deeply? Tito said that he had never met another man as loving and tender as Jim. He never imagined that he would be in a relationship with another man, let alone the depth of love he felt for Jim. Jim assured him that he had only been with a couple of other men, but none of them showed him that love is more powerful than sex, and the only concern that Jim had now was to be as good as Tito at expressing passion in their loving relationship.

In the morning, Jim began his duties looking over the battalion strength of the 6/56 artillery made up of 648 officers and enlisted personnel scattered from Saigon to Danang. In Jim's position as head of personnel, he had the tremendous power of being able to control or reassign personnel at his discretion. It didn't take long after he assumed his new position that the 160 personnel in headquarters battery realized that Jim and his close associates were not to be messed with, or they could find themselves in a less desirable position somewhere in the boonies of Vietnam. His influence and power by his position were well-known and respected.

The office of personnel for HHQ ran at 100% efficiency. The Adjutant, along with the commanding officer of the battalion, held Jim in the highest of esteem. One day the Adjutant called Jim into his office and told him that it would be impossible for him to function at his capacity without him and thanked Jim so much for his support and loyalty. Jim assured him that it was a team effort and that everyone in the office had contributed to the unit's success, and he was just the person who happened to be the lead as NCO (noncommissioned officer). Jim saw to it that all of his staff were cross-trained in each function within the office. He contacted Andrew at the relocation center in Cam Ran Bay and requested and received one additional recruit to be assigned to his old position as the utility staff member. The new recruit was SP4 Jenkins, who was also trained at Fort Bliss, Texas. He was a slender tall black man with a glowing personality and quickly fit in with the headquarters operations team. Jim's first assignment for Jenkins was for him to tour all of the personnel units of the 6/56th and to familiarize himself with their battery clerks and conduct a review of their operation.

Sometimes, Tito, David and Jim would head to the showers to gather after dinner. David was one of those guys that, up until his assignment in Vietnam, had never considered a man for any sexual pleasure. He enjoyed being around Tito and Jim because they were so natural together. In the evenings, when the three of them returned from the club, they would retire to the bunker where David would engage in self-satisfying sexual fantasies while Tito and Jim made love in front of him. Then all three of them would hug one another with their naked bodies interlocked even though Tito and Jim maintained a monogamous sexual relationship at all times.

David's body was absolutely beautiful. He was six-feet tall, firmly built, black hair all over his body, and he was very proud of his manhood. He was extremely macho but enjoyed showing his femininity in private when he was with Tito and Jim. The three of them would get together every so often, but he usually connected with Ricky when he actually wanted sex. David always commented on how much he enjoyed being a "power top," and Ricky, who was blonde and slight of build, would always willingly submit to his manhood.

Jim repeatedly feared that he could never rise to the level of sexual prowess Tito provided and continuously pushed himself to get better. For Jim making love to Tito was to engage in passionate lovemaking all while Tito made love to him. One day Jim could tell it finally happened by Tito's explosive body's reaction to Jim's passion. What a marvelous feeling it gave Jim; he was so proud of being an equal partner to the person that he was going to spend the rest of his life with. From that moment on, week after week, they shared passionate love equally with one another. Jim thought that this was the most beautiful period of his life. Not only did he feel good about himself and what he was doing, but he dearly loved who he was doing it with. Tito wasn't just any ordinary man; he was put in Jim's life for a reason, and at that time, he didn't know what it was.

It was the end of February, and Jenkins, who lived in the hooch next door, had joined into the daily routine with the rest of the office staff. When he joined Tito and Jim in the shower, they both commented to each other about his smooth, beautiful black body. He was almost hairless except around his pubic area, and he enjoyed showing off his masculinity that hung proudly between his legs. Jenkins soon began joining all of the mates at the club for a few drinks at night and sang along with David with his high-pitched voice. David commented as to how well Jenkins could sing and enjoyed having him join in on various songs.

When Jenkins got back from his review of all the battalion's line units, Jim sat him down and asked him his opinion as to the unit's personnel operating condition. Jenkins replied that all units except for "D" battery, located on the island, were adequately staffed, and records were maintained correctly. He went on to state that the First Sergeant at Delta battery appeared to be a problem. While he was visiting with Ron, the battery clerk, he was reminded of "Mouse's" attempt to blow up the First Sergeant a few weeks earlier. He also learned from Ron that the Blackburn wanted to remain in Delta battery when he extended for his third tour of duty in Vietnam. Jenkins said that Ron and the other personnel of the unit had discussed the fact that it would not be in the unit's best interest for the First Sergeant to stay in the unit for another year. His disciplinary attitude and paranoia were causing hardships to the daily operations of all soldiers assigned to the island. Just about every solider he interacted with had a negative reaction to the First Sergeant's intention to reenlist and remain on the island for another year.

Jim thanked Jenkins for his assessment concerning First Sgt. Blackburn's negative effect on Delta battery. He told David, who handled enlisted records, to pull Blackburn's file and bring it to him. Jim made an in-depth review of his file and noticed that his request for a third tour to remain in Vietnam had not been received yet. He took the file along with Jenkin's written review of Delta battery to the Adjutant for further discussion. Since they did not have any request for an extension from Blackburn, they immediately cut orders for him to report to the relocation center at Fort Bliss, Texas, for further assignment to a unit at the center's discretion. In other words, bye-bye Sarge.

When First Sgt. Blackburn received his orders, he immediately demanded a meeting with the commanding officer of the battalion. The Adjutant, along with Jim, met with the Colonel a day before his meeting and went over all of the issues that prompted their decision to decline the First Sergeant's request for a third tour and to reassign him back to Fort bliss. The Colonel concurred with their decision and was reminded of the court-martial of Mickey Jones, who was found guilty of attempting to murder Blackburn a few weeks earlier. Off the record, they discussed that perhaps there was a reason that Jones' stunt to get rid of the First Sergeant was shared by more men on the island than just him. When the Colonel met with Blackburn, he was informed that the decision for his reassignment back to Fort Bliss came from him, and he made it in the best interest of the battalion. Blackburn had no choice but to accept reassignment back to Fort Bliss.

It was now mid-March, and it was a sad day for everyone in the squad. It was time for Ricky to be reassigned back to the states and discharged from the army. This meant their beloved hooch mates David and Ricky's love life would come to an end. All of them will truly miss him. The evening before he left, they had a huge party at the club, which went on into the wee hours of the morning, and needless to say, they all drunk beyond their limits. When Ricky and David got back to the hooch, they retreated to the bunker for one last night of sexual bliss, not to be seen again until noon the next day. That afternoon they bid Ricky farewell and assured him that they would all be in touch with him when he got home and that he better writes back.

After Ricky's departure, there was room for Jenkins to move into the hooch. He was a welcome addition to the collection of personalities that ate, showered, partied, and sat around evenings sharing their personal lives. It didn't take Jenkins long to adapt to the routine comfortably. In a way, he was a blessing in disguise. Matthew and Andre's relationship had blossomed into an obvious lovefest. They no longer retreated to a remote area in the compound to share their affection, but now openly displayed it. Tito and I were glad to see that their relationship had matured to a level to which they are now able to share it in front of their mates nightly.

David and Jenkins began showering together on several occasions. At night when we went to the club, we all noticed that Jenkins sat at the right hand of David as he sang and played his guitar. Jenkins never hesitated to participate in the songs that David sang, often putting his arm around David's shoulder, acting out the emotion of a particular song. One night when Tito and I were in the bunker together, we were surprised when David and Jenkins entered the bunker and began exploring each other's sexuality. This was the first time that Tito and I had ever noticed David being dominated as a bottom. Jenkin's tall and lean body enveloped David as he groaned with pleasure. Jenkins was so tender when he made love to David, he literally opened himself up to the bliss of affection that he had never received when he was with Ricky. David's exploration in lovemaking was a surprise to Tito and Jim as he was so macho outside of the hooch.

Jim and Tito were also making plans to take their R&R (rest & recuperation) leave together on the 15th of April. They decided to go to Australia. In preparation for the trip, Jim arranged for the utility employee from the brigade headquarters to fill his position during the week he was gone. Jenkins was fully cross-trained along with the rest of the office staff to absorb the duties that Tito would be vacating at the time.

When Tito and Jim arrived in Sydney, they checked themselves into the old Australian Hotel. It was a stately relic of the Sydney of years past. The first floor contained a pillared core door with various reception rooms, in addition to the Winter Garden famous for its morning and afternoon teas, light lunches and theater suppers. The Moorish lounge leading to a huge dining room called the Emperor Room was exquisite with its highly decorated ceiling with Italian chandeliers, a white marble fountain and other statues engulfed in Palm Court style shrubbery. The Australian became the place to stay and be seen by the upper echelons of society.

Their room was beautifully appointed. It had two beds, a large sitting area and a beautiful restroom. The first thing that Tito and Jim did was draw a hot bath and playfully soak away the ten hours they traveled that day. They were now so clean that when they fell on the bed together, their bodies squeaked when they interlocked. They were trying to remember when they had ever felt so free to make love in their life. Neither could recall a time, and both said since we are here right now, let's make this last all night. Whatever one's mind can imagine happened during that night. When midmorning arrived, they were well-rested and ready for breakfast. The room service menu was reviewed, a phone call was made, and soon thereafter, a knock on the door was answered, and the room service boy was there with their breakfast.

Their first day in Sydney started on a Monday afternoon. No plans were made for the day or evening; they just followed their instincts around town. Soon they noticed many people going into pubs, and it was only around five o'clock in the afternoon. Tito and Jim followed suit and went in and out of three different pubs and found all of them packed, shoulder to shoulder, with many young office workers fully engaged in the sport of drinking. Tito and Jim joined right in, and at about 10:30 p.m., they were utterly intoxicated and headed back to the hotel. During their stay, Tito and Jim toured all of the local tourist destinations, including the Opera House, Circular Quay port, the Taronga Zoo, Chinatown as well as a ferry ride around the harbor, which allowed them to visit various other Sydney destinations.

Tito and Jim considered their week in Sydney as their honeymoon. All of their free time together was devoted to exploring the passions they held for one another. Many hours were spent in bed, holding each other tightly, and talking in detail about their future life together. On their last night in Sydney, they came back to the hotel early. They ran a hot "bubble bath" and made love watching the bubbles burst over each other's silky bodies. After they got out of the tub, their skin was completely wrinkled from the hot bubbly water. They put on the hotel's large terrycloth robes, with nothing underneath, and ordered room service as their last dinner. Neither one of them knew that in only a few days, their hopes and dreams for the future would end and their lives as they knew it would be altered forever.

When Tito and Jim returned to the unit, their hooch mates were excited to hear about their trip to

Sydney. So that night at the club, they all gathered around to hear them tell of their adventures during their honeymoon in Sydney. Of course, David was there to serenade them all with a few songs with Jenkins joining in to complete the melody. That night Jim and Tito returned to the hooch as they were both completely drained from their trip and needed to get some rest before they started back to work the next day. David and Jenkins took full advantage of the sleeping bunker. This time David pursued Jenkin's manhole with such gusto that Jenkins could not help but moan and pant with such intensity that everyone could hear him as if he was right beside them inside the hooch.

When Jim arrived at the office the next morning, the first thing he did was contact Walter, the utility employee from the battalion, who covered his position while he was in Sydney. Walter was pleased to inform him that everything ran smoothly, and everyone got along beautifully. He informed Jim that he stayed in his bunk during the week that he was working in the office and that the only thing that he had observed was how very close and kind all of his roommates in the hooch were. Jim's comment to that statement was that they are more than just brothers at arms; they are his family. We work together, we eat together, we shower together, we go to the club together, and we operate as one team to get the job done daily.

After Jim got back his trip to Sydney, he took the time to reflect upon the closeness that he had shared with Tito. Tito had taught him to truly and completely accept himself. The loving acts he committed with another man were not only a beautiful part of life but also totally acceptable. Jim understood that this increase in self beauty and acceptance could be used as a secret weapon in his personal and professional life. That special glow/ sparkle that now showed through him when he walked into a room could be attributed to Tito's love radiating through him to the world.

MOONLIT

*B*efore May 4^{th,} our unit had sustained a few rocket and mortar attacks; each time, we would proceed to our bunker as quickly as possible and stayed until the attack subsided. On the night of the fourth, David, Jenkins, Tito and Jim had showered and joined their mates in the hooch. All six of them spent the evening surrounding David and Jenkins as they sang songs, until about midnight. It was one of those rare nights that they did not go to the club and decided just to wind down to have a more peaceful night in. After David's last guitar strum, they all decided to head for their bunks to go to bed. After they wished each other good night and the occasional I love you, they all headed to bed.

2:07 AM that morning, everyone awakened from a gigantic boom that shook their beds and knocking nearly everyone to the ground. Dust and smoke softly blanketed the room, and for a brief moment, all was still. A rocket had made a direct hit on the southeast corner of the hooch where Tito and David were sleeping. The rocket hit with such force; it collapsed nearly a fourth of the hooch. Jim slowly raised himself off the ground as sand and debris fell from off his body, his ears ringing from the concussive blast, he stood up and took a step stumbling a bit, and began to look around. The moonlight gently shone in from the blast zone the rocket created where he saw Tito lying lifeless on his bed. David was yelling in pure agony on the ground near the sleeping area next to Tito. Without a moment's hesitation, Jim told Jenkins to bring the medics to the hooch as soon as possible.

Jim's attention turned immediately to David, whose wounds were massive and gushing and was bleeding extensively from the lower half of his body. He and Matthew put David on his bed and pulled him close to the back door just in case the ceiling collapsed.

Jim bent over David and kissed him and told him that he would never leave him. David took one arm and pulled Jim down to his face and kissed him back, and said, "Thank you." Just then, the ambulance, along with all the battalion medics, arrived to attend to David. Jim asked one of them to check on Tito. As David was being prepared for transit to the base hospital, the medic returned to tell Jim that Tito was indeed dead. Jim was in shock and couldn't believe what he just heard. Jim slowly walked to where Tito's body laid, tears filling his eyes. Tito's lower body was full of shrapnel and covered in blood. It did not seem real; his body was still warm. Jim took his poncho liner and covered him up from his neck down. He bent over and kissed Tito's forehead one last time, smelling the fresh citrus of the soap from the shower that they had taken only hours ago. In his mind, Tito smelt alive.

Jim was paralyzed. He refused to believe what just happened. Tears traveled down his face, falling onto his lover, cleansing him of any imperfections on his angel-like face. He stood there over Tito as if time paused, not moving an inch until the medic and his aid returned to the hooch to attend to Tito. The medic knew that Jim and Tito were more than friends, and he asked Matthew and Andre to take Jim outside the hooch while they prepared Tito's body for transport. Within minutes two people in Jim's life had been taken away. Matthew and Andre had their arms around Jim, attempting to hold back their tears. Soon Jenkins joined them, totally devastated by the events. He said the medics had told him that David would only be temporarily in the hospital at division because his wounds were so severe, he would need immediate transportation to Japan for further medical treatment.

As morning approached, all four of them had returned to the shower to clean off the blood from their bodies and to get ready for work that day. Jim was on autopilot with his emotions buried deep within. Jenkins was slow to perform his duties and would often stop and shed a few tears before he resumed his work. The entire staff of the office had gone through a life-changing event that could only be described as catastrophic. For Jim, he knew it was the end of his life as he once knew it, and at this very moment, he did not know how much longer he could continue living. The night after the rocket attack, Jim and Jenkins stayed locked away in their bunker, crying what was left of their hearts out while attempting to comfort each other's sorrows and pain.

It was during this period that Jim realized that his entire life could be summed up in months, weeks, days and hours. Tito had been like life's finishing school for Jim. He taught Jim how to appreciate his inward beauty and to accept himself as a gracious man. He shared with Jim a view of himself that he himself couldn't see, showing him the strength and poise, he commanded as he walked into a room while all of its occupants focused their glare upon him and listened intensely to what he said. Time and time again, Tito would anxiously predict their future together and dream of working at his father's company in Miami.

Jim humbly and passionately requested that the Battalion Commander allow him to escort Tito's body back to Miami. He knew it was a long shot, for he was sorely needed at his post. To his surprise, his request was approved. Brigade headquarters then made transportation for Jim to escort Tito's body back to Miami. Jim telephoned Tito's father, Henry Martinez, to introduce himself and inform him that he would be accompanying his son's body back to Miami and would be arriving within the week. Jim had no idea if his parents knew anything about their son's relationship with him, and he gave serious thought on how to handle the entire situation without breaking down.

When Jim arrived in Miami, he escorted Tito's body to the designated funeral home, where he met his parents Henry and Mary Martinez. His parents were absolutely the most delightful people you could ever meet. Jim found out that Tito was their only son and that he was an outstanding athlete and a state champion tennis player. His mother informed Jim that Tito had always been a very sensitive and loving child and looked for the good in every individual that he met.

After everything was finished at the funeral home, Mr. and Mrs. Martinez asked Jim to join them at their home for the remainder of his stay in Miami. Jim graciously accepted the invitation and was amazed to see their residence. Apparently, Mr. Martinez's business was a very lucrative one for his estate was located on an island known to house only the Rich and Famous. Driving down his street, Mr. Martine pointed out random mansions telling Jim of the business owners, film producers, and other celebrities that lived there. Outside their home, they had a private dock containing a fifty-foot yacht tied up bearing the name "Tito." On one side of their house was an enormous Olympic size pool and on the other side a full tennis court.

The Martinez's home was quite large and had numerous bedrooms, but Mrs. Martinez suggested that if Jim wanted, that he could stay in Tito's room while he was their guest. Jim accepted the invitation without hesitation because he could be close to the relics of Tito's life for the last time. After Jim was settled, he showered, changed into his civilian clothes, and joined the Martinez's on their lanai for cocktails. During several cocktails, Henry proceeded to tell Jim that he had received several letters from Tito telling him all about his relationship and plans for Jim to join the family business after Vietnam. Jim was surprised to hear the calmness and acceptance in his voice. Tito's mother smiled warmly at Jim and turned to her husband and said, if Tito loved Jim, then she would love Jim as well. To her knowledge, Tito had never been in love before, and if he loved you, she knew that Tito intended for his love to be everlasting. Jim had never been around such understanding and loving people in all of his life, and he felt utterly at home with the Martinez's.

For the next three nights, while he was at the Martinez's home, he smothered himself in the blankets on Tito's bed and smelled all of the clothes in his closet. He read all of Tito's diaries that he had left behind, and none of them revealed any love for any men, only for the sport of tennis. His writings did reveal that he was very lonely and had been searching for love but had yet to find it. The last entry in the diary was the date that he left for the army. He expressed no fears, only that of excitement in meeting new challenges in the days ahead.

The family buried Tito in a stately mausoleum that also held relatives of the Martinez's. The funeral was held with full military honors and went off without a hitch. The week that Jim dreaded turned out to be a week of celebration of Tito's life. Hundreds of Tito's friends and family members showed up at the funeral, and accolades were spoken to the highest degree that anyone could bestow upon an individual. All of the individuals that spoke were young men and his tennis coach that he had for eight years. He was so highly regarded in his world of tennis that it was hard for Jim to reconcile how he fit into Tito's life.

A day before Jim was to return to Vietnam, Mr. Martinez took Jim to his import/export office located in the port of Miami. He took Jim into his private office, which on the wall exhibited pictures of ships, planes, and office locations in various ports around the Caribbean Sea. Mr. Martinez told Jim that his operation was global and massive, and it takes a well-organized individual to run it. "Tito told me you were the man to do it," Mr. Martinez stated. He also informed Jim that Tito's heart was not really into the family business, but he was good at public relations, not so much operations, and that would be Jim's job if he wanted it, that is. This, of course, is all pending on when he is discharged from the army. Jim thanked him and that he would consider the opportunity, but at this time, he cannot make any promises or decisions of that magnitude because he must return to Vietnam to finish his tour of duty. Jim thought to himself that everything Tito had discussed with him about his future with his dad's company was true and all the detailed planning that he previously discussed with Jim almost nightly had already been put into motion.

RETURN TO VIETNAM

The next day after Jim touched down and returned to work, he was notified that the entire battalion was going to be deactivated and returned to Fort Bliss, Texas. It was the first unit in Vietnam that President Nixon was going to withdraw as part of his overall plan to downsize troop levels in Vietnam. Jim, as head of personnel, was given an order that the entire battalion of 648 men needed to be ready to be flown out of Chu Lia within sixty to ninety days. This was a massive task for any group of individuals to pull off, let alone the Army's Golden Boy. This was an enormous challenge for Jim to focus the entire team in one direction while keeping both their minds and his off Tito and David's loss for the time being.

Jim reorganized assignments for Jenkins, Matthew and Andrew, making them in charge of both enlisted, officer and finance records respectively for each unit of the battalion. They issued orders to relocate line batteries from their remote locations to be housed at staging areas at Headquarters Battery. When all of the reassignments were completed, there should be a total headcount of 648 men in one location, ready for an unannounced departure date and time out of Vietnam. Jim and his staff worked sixteen hours a day for over a month in preparation for the unannounced departure date.

During this time, Jim buried his emotions in the work that he was completing. The bright light at the end of the tunnel was that he and Jenkins would leave Vietnam on the command flight to Fort Bliss, as they were charged with the reassignment of all personnel in the battalion.

The end of July couldn't come soon enough. They finally received a secret communication that nine C-141 Starlite aircraft would be arriving on a specified date, at two-hour intervals and to have the troops readied for boarding at the airport at specified times. Jim's team prepared manifests for each flight and boxed all of the personnel files for each individual on each flight accordingly. Further instructions were given to the first sergeants and the commanding officers of each unit as to their boarding instructions, and transportation was arranged to transport them to the airport.

One day before the anticipated departure date, a mandatory battalion formation was held on the compound. The Commanding Officer for the battalion requested that Jim attend the formation, which generally Jim never did. During the formation, Jim was awarded several citations, including the Bronze Star and the Army Accommodations Medal. Jim's peers in the formation yelled out "hip-hip-hooray" several times and singing a chorus of "he's a jolly good fellow" to Jim's surprise. All of these events took place while the Commanding Officer was standing directly in front of Jim shockingly enough. When all of the accolades were finished, he shook Jim's hand and thanked him for the contribution that he made to not only the battalion but to every individual in it. The CO said that Jim had been an inspiration to the troops by his high performance in completing his duties and kindness to his fellow soldiers. Jim was eternally grateful and saluted the Commanding Officer, and after he turned to the entire battalion, with a tear in his eye, gave all of them a long solid salute.

The flight back to Fort bliss took 20 hours, with a brief stop in Guam for refueling. Jim sadly left Matthew and Andre behind along with the Adjutant to clean up the final reassignment of equipment, and the last-minute soldiers that were in the pipeline to be assigned to a new unit had to inform them that their orders could not be changed. Matthew or Andre would have to find a new assignment for the remaining time to new units in Vietnam. Jim took Jenkins with him to Fort bliss for two reasons. First, Jim would need help processing the 648 men and, second, Jenkins was emotionally drained after the loss of David, and he felt it best that Jenkins return stateside to serve the remainder of his time in service at Fort Bliss.

It took five days to process the 648 men reassigned from Vietnam to Fort bliss and other Army bases in the United States. Jim was granted a complete discharge from the army because, by this time, he had less than thirty days to serve before his scheduled release date. He saw to it that Jenkins was assigned back to Command Headquarters, where he was previously trained before going to Vietnam. Now Jim was faced with the ultimate decision, reenlist, or go home. There was nothing left for him in the

Army; Tito was gone. There was nothing left for him outside the Army because Tito was gone. Jim was lost and didn't know where or what to do. Jim gave thought to Mr. Martinez's offer to work for his company in Miami. But he didn't know how he could live in Tito's shadow for the rest of his life, constantly being reminded of his loss at every turn. Jim thought it would probably be too much for him to bear, so he contacted Henry and informed him that he was not going to accept the position and that he was going to return to his former position at the bank in Illinois. Mr. Martinez understood the reason why and told Jim that he would always be there for him if he needed anything, just call.

Before Jim left Fort Bliss, he and Jenkins got together for a couple of days to reminisce on the good times they had spent with their lovers in Vietnam. During the evenings, they would go out to the gay bar in El Paso and have a few drinks and venture off to a few other locations to meet up with some of the other gay soldiers from the former 6/56th. To their surprise, while they were sitting at the table at the end of the bar having a quiet conversation, Ron from Delta Battery walked in. Jim and Jenkins were both shocked to see him there. They had both discussed how much fun he was, but they never suspected that he was gay. They summoned Ron to join them at their table and bought him a drink.

Ron said he wasn't surprised in the least to see Jim and Jenkins at the bar. Ron said they had resisted all temptation in his attempts to seduce both of them while they visited the island. He said that now was his last chance, so drink up. Jenkins and Jim both laughed and challenged Ron to a drinking contest to see who would outlast who. Ron was soon "overserved," and since Jenkins and Jim were staying at a hotel off base, they took Ron with them back to the hotel. This was the first time that either Jenkins or Jim had sex since Vietnam. So that night, the three of them drunkenly explored their sexuality until the morning hours arrived.

It was now time for Jim to say goodbye to his friends. He had decided to leave the army and return to Illinois to work and move in with an old childhood friend named Gary. Jenkins lived in Oklahoma and planned on returning there, for he felt that the military life did not suit him anymore, while Ron planned on returning to Michigan after he gets out. Jim had every intention to maintain contact with both Jenkins and Ron, of course, when time permitted it. Ron got out of the Army in October 1969, after which Jim made an overnight trip to Michigan to visit him. During the brief visit, he discovered that Ron was engaged to a woman who he was now living with, whom he planned to marry sometime early next year. He wished Ron the best, and on his drive home to Champaign, he reflected on the life-altering events that he had experienced in Vietnam. He learned how to love and be loved, how to accept himself as a gay man, and that love is more powerful than just sex.

LIFE AFTER LOVE

*J*im went back to work at the bank in Champaign. They welcomed him back with open arms, but everything felt different, and the bank seemed even smaller than it was. The people that surrounded him noticed a magnificent change in him and were curious about his time in Vietnam. Now there was nothing he wanted to share with anyone, especially about losing Tito. He felt that if he worked hard and had several drinks after work, he could blot out the loss of Tito. So that was his plan to cope with his loss for the foreseeable weeks or months ahead.

One day the president of the bank called Jim into his office. He politely told him that he was wasting his talents at his small bank. He told Jim that his personality, looks and communication skills (something to that effect) were so outstanding that he should pursue opportunities at a larger bank that could offer him a career worthy of his many talents. Jim was amazed, but he agreed with the president's assessment of the limited opportunities at his small bank, where he would never achieve the earning power he needed or wanted to attain his financial freedom.

When he got back to his apartment that night, he went to his minibar and poured his usual drink of scotch and pondered over the discussion he had with the bank president. He knew that his life with Tito would have been successful and happy. Now, alone in the world, he could only measure his degree of happiness by the wealth he can attain through hard work. Without any formal education, such as a college degree, he would have to rely upon his PLC skills to reach his ultimate goal. He called his grandmother, who he knew had a cousin, of some wealth, in California. He inquired as to her name and phone number because he wanted to ask her for some help with his job search.

He called Delpha, his grandmother's cousin, who lived in Bellflower, California, and told her he wanted to interview with one of the large banks out there. During their conversation, she stated that she banked with the world's largest bank, and she would be happy to make a phone call on his behalf. Apparently, the bank respected her and her wealth because Jim received a phone call from the regional manager asking him to contact the training officer to make arrangements for an interview, at his convenience. Without wasting any time, he contacted Tim, the regional training officer, and made an appointment the following week to meet him in Santa Ana, California.

Jim arranged for a flight and a car rental; he stopped by and visited with Delpha, who was so kind to connect him with the bank before he went to his interview with Tim. The interview lasted about thirty minutes. Then Tim gave Jim a letter indicating that he could start immediately with an officer training program at the bank. Jim returned for a short visit with Delpha, thanking her again for her assistance, before heading to the airport to catch his flight home. All in one day, Jim had flown to California and secured a position with the world's largest bank. That evening he returned home and now began to plan for a new beginning in California, finishing his day savoring a stiff scotch from his minibar.

While drinking his fourth, Jim reclined in his chair and closed his eyes to reflect on the time that he and Tito had together. Tito had made Jim feel so good about himself, and with that feeling, he could manage any or all events that he faced in life. Tito had always commented on Jim's personality as one that fits every occasion and that his looks were TV ready. His communication skills, Tito said, were what set Jim apart and that he was a great listener who always knew what to say at the right time. Tito often spoke of how clear, concise, and correct Jim was when he addressed people as a group or individually. So, in reality, his loving relationship with Tito had reinforced Jim's self-confidence to the level that he was so self-assured he was ready to tackle or face any obstacle big or small.

His relatives, co-workers, and a select group of friends held a few going away parties for him before his leaving for California. The most touching of all the parties was that of the selected group of gay clientele that he had at the bank. They hosted a private party at one of the clients' extraordinary private estates. The Who's Who of the gay elite in the Champagne area gathered to recognize Jim's service to their community and assured him that he would be missed not only at the bank but at their gatherings.

THE ROAD TO CALIFORNIA

*J*im loaded all of his worldly possessions in his car and asked his roommate Gary if he was able to ride along with him to California. Gary said he had friends in San Diego and that he would enjoy visiting them. Jim said he could reroute his trip to go to San Diego if he would join him on the long journey ahead. He agreed, and they took turns driving nonstop all the way to San Diego, where Gary joined his friends. Jim headed north to Bellflower to revisit his cousin Delpha who offered to temporarily house Jim until he could find a place of his own. While he was with Delpha, he located a small studio apartment in Long Beach, where he could stay during his training program.

Jim joined the bank on May 1, 1970. The operations officer training program was to last twelve months. He found that the training program to be very similar to that of the Army's, so completing it in six months was not a challenge. Additionally, he completed two bank operations assignments very successfully during the next year. Then an opportunity came up for an international assignment as an operational analyst. Jim quickly applied. He was relatively surprised that he got the position for he did not have a college education, but what he did have was an outstanding Army career as well as an equally exceptional referral by the president of the bank in Champaign, Illinois. Jim's felt that his career had now taken off, and nothing is going to stop him from achieving his goals. Jim worked nonstop, continually challenging himself to be better and better every day while always keeping in mind the end result that he wanted for his life's quest.

During approximately ten years of working internationally, he was recognized as the go-to person for expertise in international bank operations. He interacted on numerous occasions with Lee, the Chief Financial Officer of the corporation, who managed international banking and advised him on many operational issues. His opinions and recommendations were almost always looked upon as a standard when it came to international operations.

He was amazed at how easy it was to conduct these internal compliance audits and how similar they were to the Army's rules and regulations. He traveled through fourteen countries throughout the world, many of them two or three times during his ten years.

While in these countries, Jim would usually retain a full-time butler who would be at his disposal 24/7. He would escort Jim to various restaurants, nightclubs, or shopping outlets, depending on Jim's wants and desires on any particular day. Often these individuals would comply with Jim's sexual desires. Throughout the years, Jim did not develop any emotional relationships with any man in any of the various countries. In fact, Jim always felt that his heart must be chiseled out of the coldest of stones, and he was all business.

In the last year of his decade of international assignments, he was asked by Lee to conduct a few motivational seminars covering the "Vision, Values and Strategies" of the corporation. Jim stepped up to the challenge and was very successful at it. His outstanding communication skills, along with his personality, ranked him as the top speaker of the corporation. The majority of his classes were held in Manila in the Philippines; the participants came from banking units all over the Asian and South American Divisions. While in the Philippines, Jim lived at the Manila Hotel and his suite was next door to the suite maintained by General MacArthur, who lived there during World War II.

During the weekends/weeks when he wasn't working, he would travel to various countries that he had not worked in to meet the people and tour their cities. Jim developed a very worldly personality, and by this time, he was starting to attain a level of social consciousness. He felt comfortable around both affluent or non-affluent people, for to him, it made no difference where you came from, as long as you had a good heart.

There would be times when he was alone, sitting at a bar having a drink, and would reflect on his past, growing up as a son of sharecroppers and how devastatingly poor he was. He was so far from that now it was unbelievable, and the twist and turns that he had been through in his life in getting to where he was today seemed amazing. Jim always thought about the happiest time of his life, when he and Tito had been together; those were his most cherished memories. He learned and accepted more about himself in those eight months with Tito than he did in all of the years that he had lived. His love affair during that time of war with Tito established a remarkable confidence in himself that would enable him to achieve all of his life's goals, both personal and financial.

Jim was now coming to the end of his ten-year international assignment. He had given Lee, the CFO, the heads up that he was not going to renew for another two-year overseas assignment. Sitting back and relaxing, Jim reflected upon the last decade of his life and the people who had come and gone. He thought of Matthew and Andre that took care of him at the Cam Ran Bay Reassignment Center; he wondered where they were at today. The last time he had talked to Andre was when he called in a request to fill Jenkins's position. He had gotten two letters from Ron, the D Battery Clerk, that he had visited in Michigan. In one of the letters, Ron explained that he had gotten a divorce and asked when Jim was coming home to Illinois. Jim thought it's about time that he contacted Ron and find out what he's been up to because it has been far too long since he had last talked to him.

Jenkins, bless his heart. He had found love in the great state of Oklahoma. We wrote back and forth two or three times a year, and he was delighted with Jared, his new lover, and they wanted to get together with Jim whenever possible. Jim thought he would look up Jenkins and Jared just for old time's sake. Jim was also very grateful for his friend Casey; the two of them had flipped three houses in California while he was overseas and made a considerable amount of money. Casey had Jim's power of attorney and acted on his behalf during the real estate deals. When the sales closed escrow, Casey would get a percentage of the net proceeds. All around, it was a good deal for them both, and they both made some serious money.

During the last two weeks before Jim left South America, he was living at the Tamanaco Hotel in Caracas, Venezuela. On weekends it is like a country club for the rich and famous. Gorgeous young Venezuelan men, clad only in Speedo's, surrounded the pool, tanning their golden-brown bodies in the rays of the tropical sun. Jim had his butler secure a lounge chair mid-way at the pool where he would have a good view of all of the Speedo-clad bodies while he ate his breakfast and lunch each day.

It was during this time that he received a phone call from Lee, the CFO. After pleasantries and small talk, Lee inquired if Jim had plans after returning from South America. Jim responded that he had not spoken to human resources about his next assignment, so at this time, his future is up in the air. Lee asked Jim if he would stop by and see him as soon as he got back to San Francisco, and no appointment was necessary.

Jim purchased his ticket to San Francisco via Tulsa, Oklahoma. He phoned Jenkins and told him that he would be there for a short visit at the end of the week. Jenkins met Jim at the airport, and they met each other with a long loving embrace. Both of them were so delighted to see each other that tears ran down their cheeks. Jenkins had taken off Thursday and Friday so that the three of them would have a long weekend together before Jim left Tulsa on Monday morning for San Francisco. When Jenkins pulled into his driveway, they were greeted by Jared, Jenkins's lover of nine years. Jared was strikingly handsome, as was Jenkins; both of them were black and very successful. Since Jenkins return to the states, he has been employed as the human resource officer for American Airlines based out of Tulsa, and Jared was a CPA for a major accounting firm in town.

Jenkins and his partner had purchased a beautiful two-story home in an upper-middle-class suburb of Tulsa. Its architectural design was colonial, and it had four bedrooms and five bathrooms. It was quite obvious to Jim that they were in a loving relationship and enjoyed each other's company to the fullest. For the next three days Jim and Jenkins reminisced about old times they shared in Vietnam. Over the next three nights, the three of them went out to some of the best restaurants in Tulsa, where Jim was delighted to pick up the tab. On Sunday night before Jim left for San Francisco, Jenkins and Jared prepared a meal at their home that outdid all of the restaurants that they had visited by far. Everything was cooked with love and beautifully dressed. Not a detail was missed by the two, and Jim wondered why they were not chefs. That evening after dinner, they enjoyed a few cocktails in the mahogany-paneled library of their home when Jenkins got up out of his chair and sat on the sofa right next to Jim. He laid his head on Jim's shoulder, and small tears ran down his cheek. He told Jared to join him on the sofa, and both of them consoled Jim with kisses on his cheek and their heads on his shoulder.

It had been a very long time since Jim had received any kind of love. He trusted Jenkins's emotions as real and genuine because Jenkins knew what Jim had gone through and how emotionally vulnerable he was. Over the years, Jim had hardened his feelings using long hours of hard work and plenty of alcohol. The less time he spent thinking of the past, he thought, the better off he was. As the clock struck midnight, Jared took Jim's hand and led Jim to their bedroom. Jim let his guard down, for once, and undressed as did the other two and got in bed between the two lovers. They spent the entire night lovingly interlocked, kissing and fondling each other until Jim exploded with an intense orgasm, like none that he had felt for years.

The next morning after a quick breakfast, Jared gave Jim a hug and a kiss and told him that he had to head out to the office, and Jenkins would take him to the airport. As Jenkins drove, Jim thanked him for a wonderful long weekend and that his partner Jared was the perfect match for him. Jim said when he got back to San Francisco that he had a meeting set up with the CFO of world operations, but at this time, he did not know where his next assignment would take him. He told Jenkins that he would be in touch with him as soon as he found out where and what he was going to be doing.

When Jim arrived in San Francisco, he checked into the Fremont Hotel, one of the best luxury hotels in the city. It had a world-renowned reputation for impeccable service, and its central location was convenient for his meeting with Lee at world headquarters. That evening Jim enjoyed an exceptional dinner at the Tonga Room and iconic Hurricane Bar restaurant located within the hotel, which has served guests since 1945. The next morning at about 9 o'clock, he gave Lee a call and told him that he was in San Francisco at the Fremont hotel and would be available to meet with him at his convenience.

Lee did not hesitate in the least and asked Jim to come to his office immediately and that he would look forward to meeting with him. When Jim met with Lee, and after all of the pleasantries, Lee inquired if Jim had made any career plans after returning to the United States? Jim responded just as he did before, that he had not spoken to human resources about his next assignment, so at this time, his future was up in the air. At that moment, Lee leaned back in his chair and looked Jim in the eye and complimented him on his ten years of service in the international division. He went on to tell Jim how much he had grown to rely upon Jim's expertise in international operations, which assisted him in making many important decisions.

Lee was sixty years old, lean build with gray hair, and very soft-spoken. He informed Jim that he was planning to retire in a few years, and the international operation was becoming more challenging every day. Along with his primary duty as the chief financial officer of the corporation, management of the international operation was starting to become a drain on his mental capacity. He then said that he needed assistance from a person like Jim to make his job easier and asked Jim if he would consider being his personal attaché, responsible only to him for the international operation sector of the bank. Lee assured Jim that over the past ten years, he had closely monitored all of Jim's reports and assessments and felt comfortable that Jim could handle the assignment and keep Lee apprised of any situation that would come up.

Jim reeled back in his chair in total astonishment as to the complexity and the vastness of the assignment that Lee just presented to him. He asked Lee if he could have a day to think about it, he would like to come up with some questions as to the parameters for the job and what Lee would expect from Jim. Lee agreed to meet Jim tomorrow morning at nine o'clock in his office and told Jim that he sincerely hoped that Jim would be by his side from tomorrow forward.

On Jim's way back to the Fremont, he stopped him at one of his favorite bars located at the St. Francis Hotel. The descendants of the Crocker family constructed the St. Francis on Union Square in 1904 at the tune of $2.5 million; they intended for it to be the Paris of the West, and the grandeur of the hotel was breathtaking. Jim entered the hotel and went directly to the Clock Bar, which takes guests on a journey through time, inspired by a specific era. Jim ordered a double scotch with soda as he had much to ponder after today's discussion with Lee. While he was sitting at the bar, he thought about Ron and thought that he should give him a call when he got back to his room. He thought it was about time that they got together to rekindle their former relationship, possibly.

For any significant decision Jim ever made, he slept on it overnight. That night wasn't any different from any other night where a decision had to be made. He tossed and turned, visualizing the magnitude of the responsibility of the position, as Lee described it. He wasn't even a Vice President, so how could he hold positional power over all of Lee's direct employees. To succeed in this position, Jim would have to have ultimate control over all the incoming information and the dissemination of information for all of the international operations. The final decision would be made tomorrow morning. Before all the anticipated tossing and turning, Jim decided to talk with Ron when he suggested that he fly to San Francisco for a visit. He said he would let Ron know when he got settled in and work out the details of the trip later.

Jim arrived at the world headquarters at about eight o'clock the following morning. He asked to use an empty office where he wanted to make some notes before he met with Lee. On his list, he included the following items:

1. VP title
2. Control over incoming and outgoing information
3. Personnel in Lee's operation reported directly to him

4. Number of years Lee visualizes the assignment would last
5. His reporting relationship to other executives in world headquarters
6. The bottom line of what Lee actually expected of Jim

At nine o'clock sharp, Jim knocked to announce his arrival. Lee got up out of his chair and met Jim at his door; he told Jim to follow him and led him to a beautiful corner office facing the Golden Gate Bridge, which had a view of the entire city of San Francisco. Lee informed Jim that if he took the position, this would be his office. They both had a seat in this very lavishly appointed office, and Lee inquired as to Jim's decision.

Jim started to communicate his list of concerns about the position. Lee listened intently, and when Jim was done, Lee said everything on your list would be done. Starting today, you are a Vice President, and all of my direct reports must go through you with their concerns and recommendations before they communicate with me. Lee reminded Jim that he had already told him that he planned to retire soon, and he had not yet made up his mind when that was going to happen, but he would like Jim to stay in his position until that date arrives. He said that he would not be able to replace him and identify another replacement as it took ten years to find Jim. One additional item they covered was he was to be on call 24/7 for international operations are in different time zones in countries around the world.

A compensation package was discussed, and Jim reeled back in his chair when he heard the dollar amount. His salary was doubled along with a VP's title; how could he turn down this position? Jim informed Lee that he would take the job and start immediately.

Jim laughed to himself, how could it be that this gay man raised as a son of a sharecropper now held one of the highest positions in international operations for the world's largest bank? It all boiled down to three things; his glowing personality, TV-ready looks, and polished communication skills, which he had perfected from an early age. His next challenge was to find an apartment in San Francisco that would be an easy walk to world headquarters. It took a couple of days, but he located an apartment on the 19th floor of the Fox Plaza on the corner of Market and Polk. The view from his apartment was almost identical to his office and showed San Francisco in all its beauty.

He gave Ron a call and gave him the address of his new apartment and told him that he would send him an open-ended round-trip ticket to the city so he could come to visit anytime he wanted. Ron was so excited that he told Jim that he would catch the next flight out or within the next one or two days. Jim told him that he just secured his apartment and that he would have to go furniture shopping, so give him a couple of days to get everything furnished, and he would be looking forward to seeing him.

THE EXECUTIVE

The first week at his new position was very challenging. He started to interview personnel for his secretarial position, trying to find the exact qualities that made him successful in another man. It also took Jim a couple of days to develop an organizational chart that would direct all of Lee's direct reports and all other communications to be sent to Jim, where he would filter them and pass them onto Lee. By the end of the week, he had hired a secretary by the name of David; he was twenty-two years old and had just earned his associate of arts degree in English from San Francisco State University. Dave's benefits package included a substantial salary and all schooling expenses through his Ph.D. as long as he worked for Jim.

Jim met with all of Lee's direct reports and the remainder of the office staff at world headquarters.

They had no problem communicating their issues to Jim before discussing them with Lee. Almost all of them knew by way of reputation who Jim was, which made for a smooth transition. They were also informed that David spoke on Jim's behalf in his absence.

One of the first personal things that Jim had to do was to purchase an entirely new wardrobe. His shoes, shirts, and suits revealed miles of travel and were worn out. On his way home from his first day of work, he stopped at Brooks Brothers and invested over $10,000 in new business attire. Jim also gave thought to the furnishing of his new apartment at Fox Plaza and started by purchasing a queen-size bed that had delivered the day before.

To Jim's surprise, he got a phone call about midnight from Ron informing him that he was at the San Francisco airport and wanted the exact address of Jim's apartment. With excitement in Ron's voice, he told Jim that he couldn't wait to see him. When Ron arrived at Jim's apartment, they both embraced each other with a long hug, and they picked up where they left off when they were together in El Paso. Jim asked how long Ron could stay with him, and Ron said he had no plans of returning to Michigan because it was now time to start over with his new lifestyle. Jim gave serious thought to what Ron said but told him that they would need to get in bed soon because he has to go to work early the next day, that they would discuss his plans further after he got off work that night.

Jim's work schedule entailed him arriving at the office at approximately six o'clock every morning to study the daily "Red Bag" that contained all of the consolidated paper communication from all the units around the world. He would prioritize all of the documents in order of importance in preparation for Lee's review when he arrived at the office at eight o'clock. If there were any serious issues, Jim would research them and include a recommended solution attached to the document for Lee's review. Jim would usually get home from work between six and six-thirty.

After Jim got home from work that day, he and Ron went furniture shopping. They outfitted the apartment with a beautiful selection of leather furniture complemented by stainless steel tables and chairs. Jim asked Ron if he would finish furnishing the apartment and gave him a credit card to pay for all of the sundry items.

They then went down to one of the bars in the Castro area of San Francisco and had a few cocktails to relax. After a couple of drinks, Jim asked what Ron's plans were for the future. Ron said that his past had been so traumatic since leaving Vietnam. He had made some regrettable decisions and that he had been afraid to come out of the closet. He wanted to live here in San Francisco so that he could comfortably be a gay man and have a gay lifestyle. He went on to say that he had always been attracted to Jim and wondered if they could have any type of relationship together?

Jim thought about his past life with Ron. He had always enjoyed being around him, and he found him very attractive because of his comical attitude and manly appearance. He was one of these guys that you could take anywhere because he didn't look particularly gay. In fact, he had kind of a backwoods midwestern sex appeal, which Jim found unusually sexy. Ron's looks would never compare to Tito's, but no one could. However, Ron's kindness and understanding were a valued asset that Jim liked. Jim and Ron had not really had full-on sex together, so Jim did not know what their chemistry would be like in the future. Jim did not want a one-night fling or short relationship, but he had career goals that he was going to achieve with or without another individual at his side, and whoever it was must complement him and not distract him from his life's quest.

That night they went out for Chinese food at one of the many restaurants in Castro, the gay area of San Francisco. Jim watched as Ron scanned the restaurant looking at all of the good-looking men that were his competition. He just laughed and said, "Gee Jim, this is like a smorgasbord for you, isn't it"? Jim replied, "I rarely go to the Castro, but I thought we would tonight just for your enjoyment." Ron quickly came back with a reply, "Jim, you are all I've ever wanted." Jim was taken back by his comment.

Jim had often talked with Ron since Vietnam, and Ron had never mentioned having an attraction to him. Jim asked when he first started having these feelings. Ron said the very first time he met Jim on the island in Vietnam. He said that Jim was so strikingly handsome and in command of everything and that he was sure that there were many other gay and straight men that were attracted to him that Jim probably didn't even know. He went on to say that Jim's co-workers in the office, in Vietnam, had casually mentioned that Jim had been in a relationship with Tito and that he was devastated when Tito was killed.

When they got back to Jim's apartment, they had sex together for the first time. Ron took the lead as he had years of pent up anxiety, waiting to fill Jim and gorge upon his flesh. Knowing that Ron was once married, Jim is not surprised that foreplay was a big part of his sexual moves. He used his hands and lips to stimulate Jim in all areas of his body, leaving none untouched. Again, like Jim did with Tito, he laid back and opened himself to receive whatever style of love Ron wanted to bestow upon him. Ron was incredibly soft and gentle; he had an almost magically smooth body with lush blonde pubic hairs. Between his long muscular legs, his sizeable uncut manhood slowly rose to action. He was an incredible lover and took full charge of delivering his long-overdue passions. That night both Ron and Jim were completely sexually satisfied for the first time for many years. Jim had renewed his sexual self-confidence to the level it was while he was in Vietnam. Could it be that Jim found his new partner in Ron?

SAN FRANCISCO IN THE '80s

*T*wo years passed, and since Jim was growing impatient with Lee, he asked him as to his plans for retirement. Lee said that he was still unsure of the exact date and asked Jim to be patient and to stay by his side until that date arrived. Lee adored Jim; their relationship was more profound than a boss-employee relationship. During the past two years, Jim would always have lunch with Lee whenever he was in his office.

Before Ron's arrival, Jim's sex life was almost nonexistent. And it had been an even longer time since Jim had felt the snug embrace of love. At about three o'clock, Jim was getting ready to receive his next banking assignment; when he suddenly had an intense rush of feelings that overwhelmed him. It was like the proverbial light bulb when he finally accepted his burning desire to develop a relationship with Ron.

Jim and Ron's apartment was only five blocks from Castro Street, the gayest part of San Francisco. Jim rarely ventured out in that area of town because of his high visibility as a bank executive. During the '80s, the AIDS epidemic was rampant in San Francisco, and being outed as gay was a high hurdle for all gay men to overcome in the workplace due to the fear and ignorance surrounding the disease, which was primarily associated with being gay. Jim had always maintained a strict corporate appearance and was never questioned about his sexuality. Jim recognized that his sexuality was one of his strongest characteristics and that many of the men his senior appreciated his PLC that radiated when he walked in the room. Jim used every resource that he had to achieve his goals.

The best hiring decision that Jim made was when he assumed his new position was the decision to hired David as his assistant. He first classified him as his secretary, but as the years progressed and his experience level increased, he was reclassified as Jim's assistant. David had exceptional talent and good looks. He could be trusted entirely with all confidential matters, including Jim's personal life, and was at Jim's beck and call 24/7. Jim opened and paid for an account for David at Brooks Brothers, so David would always look the part of a successful banker. Jim knew that he had a responsibility to take care of David's career with the bank long after Lee would retire, so any investment today was for his future.

Jim and Ron's monogamous relationship was extremely sexual and satisfying. Ron's funny personality relieved a lot of Jim's stress when he would arrive at home after work. They would enjoy a few drinks together and then have a nice dinner that Ron would prepare. Each night, they would settle into their beautiful San Francisco apartment overlooking the city, never growing tired of having each other wrapped in their arms before they retired for the night. Jim had encouraged Ron to go back to school and work on the law degree that he had always wanted to achieve. They didn't need to worry about finances because Jim made enough money for both of them, and Jim's position required him to be close to work seven days a week. There were only a few times over the years that they would rent a car and go out of town for the weekend for a change of scenery.

After Jim finished his fourth year working for Lee, he once again became impatient as to when Lee was going to retire. Over lunch that day, he asked Lee again, and Lee's reply was quick. We should start looking for another position for you, and when that's done, Jim, I think I will retire. Jim thought to himself: the timeclock has begun; I better start having this discussion with David and Ron as to our future plans, so we're prepared when the time comes.

Within the week, Jim sat down with David and asked him what he would like to do when Lee retired? David replied the only thing that he ever thought about was supporting Jim in his next position no matter what it was. Jim agreed that they were a fantastic team, but he was undecided as to what he was going to do or if he's going to stay with the bank. So, he asked David to give thought to another position within the bank or someplace else that he would like to try, and he would see to it that all of the bank's resources would support him.

During the week, Jim and Ron discussed their job plans for the future. He told Ron that Lee would retire soon and that a hard decision was ahead as to what Jim wants to do with the bank or even if he wants to stay there. Jim had invested almost fifteen years there, and he was making a substantial salary with many benefits. It would be challenging for him to better himself financially in another institution anywhere in the world, so it would be a hard decision. Ron was ready to take the bar exam and become a full-fledged lawyer. Jim was so proud of Ron; words could not describe his elation and pride. He envisioned them as a "power" couple—an operational/legal powerhouse couple that would be unparalleled in the gay community. Immediately Jim thought about Henry Martinez and his business in Miami. That would be a perfect opportunity to use their combined talent in a global industry that could pay back Tito for the wisdom and encouragement that he gave Jim while they were together in Vietnam.

He talked to Ron about this idea of contacting Henry, Tito's father in Miami, and exploring the opportunity for employment in his global import/export business. Jim had not gone into any detail with Ron before about Tito's dad's business empire. Jim took the time that night and went over everything discussed with him when he was in Miami. Ron asked Jim if he was sure that he would even remember him and suggested if so, Jim should contact Henry and make arrangements for them to fly to Miami and discuss it.

The next evening Jim searched the phone books and phoned Henry in Miami. Henry was delighted to hear from Jim and asked what he had been doing all these years. Jim summarized the last fifteen years of his life working in numerous countries around the world and the global operation which he was currently supervising for the bank. Jim told Henry that he had a new partner in life that is an attorney, and that Tito had also known him while he was in Vietnam. Jim inquired as to the status of Henry's global business operation and if there was any possibility for any job opportunities for him and Ron at this time.

Henry took a big deep breath. He said this phone call couldn't come at a better time.

"I have been wondering what I was going to do; I'm ready to retire, and I don't have anyone to take over the operation. I hate thinking about selling my business that took a lifetime to build. I always wanted Tito to have it, and if you run it, Jim, it would be a gift to Tito because I know how much he loved you. I think this phone call was meant to happen. How soon can you two fly to Miami so that we could talk about some of the details of this operation? And I would like to meet Ron as well, so let's go out on the yacht and have a long weekend together."

Jim said that he and Ron would fly out on Wednesday and return on Sunday and that he would let Henry know the arrival times by Tuesday. On Monday, he told Lee that he was flying to Miami for a job interview. Jim would make arrangements with David to cover for him for three days, and if anything came up of great importance, he would only be a phone call away. Lee asked about the position, and Jim said that when he got back, that he would go over it with him.

Jim and Ron arrived at the Miami airport on Wednesday right after lunch; a private car was waiting for them and took them to the Martinez's estate. As Jim was like that the first time, Ron was amazed at the compound where the Martinez's lived. He had no idea that Tito would've come from this kind of environment because he was so humble yet private. I explained to Ron that the Martinez's are good, unassuming people; they just happen to own a very successful empire, and through the grace of God, we may become a part of it.

About two hours after we arrived, we all boarded the fifty-foot yacht named the "TITO" and headed out to the sea. At the helm, Henry asked if Jim had ever navigated a yacht before? Well, the simple answer was no, not yet. Henry had Jim take over the controls, and he said, "It's all yours; it's easier to drive than a car."

To Jim's surprise, it was. They headed out for about five hours and anchored off of a nameless isolated island. Provisions for the weekend were already on board, so there was plenty of food and drink for everyone. There was a self-service bar, a full galley, a well-appointed salon, a lovely dining room, two staterooms, and one crew quarters. There were also three full baths and one-quarter bath on board, and the back deck was all teak with lounge chairs and tables abound. Henry said Jim, you're the bartender for the cruise, and I said, "Not a problem. I make a good scotch and soda, what do your drink?"

Henry and his wife drank mimosas, and Jim and Ron drank scotch and sodas. They drank until about eight o'clock, and Mary headed off to the galley and reappeared with a selection of fine cheeses/bread/cold cuts. That night the four of them caught up on old times, shared memories of Tito, and delved into Jim and Ron's time together. Henry was very interested in Jim's experience in his global operation with the bank. He was also interested in Ron's particular specialty in his law degree. Ron informed them that he was getting ready to take the California bar. He had just finished law school and had made an asserted effort not to specialize in any one area and to keep his options open. Over the last four years, he had the luxury of not having to work at a job while he was attending law school full time, so he studied a broad range of topics. He found none of them boring. He enjoyed studying law and practicing law, and one of his greatest strengths was legal defense because of his personality and communication skills.

That night Ron and Jim stayed up until well past midnight enjoying the stars and the smell of the salt air. This was their first time on a yacht, and they found it to be the most luxurious setting they had ever been in. Everything on the boat seemed brand-new and never been used as if it was built for them. They cuddled under the stars, and Jim thought in his mind what a gift Tito had given him. His love from Vietnam had lasted for over fifteen years, and he shared the rewards of that deep love tonight with Ron. How could he ever thank Tito? The first answer that came to mind was to take care of Henry and Mary, Tito's parents, and look out after Ron.

MIRRORS

*J*im walked alone up to the port side of the deck. As a cool breeze whipped through his hair, he thought to himself, "Over a decade and a half ago, I met Juan "Tito" Martinez, the man I fell in love with at first sight, in Vietnam. He mesmerized me with his handsome male structure, his eyes that pierced my soul, his lips, I immediately wanted to kiss, and his hands… when I shook them, I knew that they could lead me to sexual bliss. For the next eight months, which seem more like eight years, he was a powerful influence that led me to believe in myself and that with my personality, looks and communication skills, I could conquer and achieve anything in life that came before me. Tito came from a position of notoriety and wealth, which enabled him to have that magical influence over others without revealing himself. His love was captivating and further enhanced by his dignified appearance that only a man of his character could possess.

"Tito, I love you."

SAILING ONWARD

That night on the yacht, Jim made love to Ron with such passion and sexual energy he could not control his own climax. The next morning Ron questioned Jim as to where all of this newfound energy came from? Jim's immediate reply was that he was in love with him and for the first time in over fifteen years, he was no longer married to the bank. From now on, Ron was going to be his first and only priority in his life. Jim learned from his past to never take for granted the person you love for you will never know how long you may have with them.

It was about nine o'clock in the morning when Ron and Jim made their appearance at the back-deck dining table. The Florida sun was already a scorching 90° degrees with an intense amount of humidity. Henry and Mary were waiting for them as they were drinking their mimosas. Henry spoke up, "What do you guys want to drink?" Ron and Jim looked at each other, and together they said in harmony, "We'll have what you're having!"

After a couple of rounds of drinks, Mary began to fill the table with an assortment of sweet rolls, bagels, fresh fruit, assorted cheese and freshly prepared omelets. Jim praised, "Mary, you are the Queen of the kitchen; where did you learn the art of kitchen wizardry?" Oh, that's easy, Mary said; all you need to know is the phone number to "The Caterers to the Yachts," they were here bright and early the previous morning before we took off to outfit the yacht with food and goodies. She sat down with a big smile on her face, re-crossed her legs, and took a sip from her mimosa.

After the delicious brunch, the towering sun forced all of them into the main salon of the yacht. Mary excused herself and retreated to her mysterious duties. Jim, Ron and Mr. Martinez settled back and began talking business. Jim said he needed a better understanding of the details within the global network. With the flick of a switch, a 36-inch TV appeared on the other side of the room. When Henry flicked another switch, several lights appeared on the board, some red, some yellow, and some green. Each light on the board indicated one of our shipping ports.

Red light:	highest
Yellow:	low
Green:	Average

The red lights reflected the highest-earning ports, the green lights were the average earning ports, and the yellow lights were the lowest-earning ports. So, with a quick review by color, one could determine the profit centers of Henry's global operation.

Henry hit yet another button, and another monitor appeared, this one showing the net revenue for each port as of today. Henry continued to hit buttons to flick through more reports. He had complete knowledge of his world operation at all times. His clientele base was one that he had harvested over the more than thirty years that he has been in business. He stated that he keeps no written record of his client's association with him, for he knows them all on a personal basis. He stated that all of the known owners within his client base are invisible to his world, and everything is done to keep it that way. Jim was amazed at how sharp Henry was and well organized while minimizing written records.

On their last night. Ron and Jim laid on the bow under the stars and discussed the secret millions of dollars Henry makes. Jim always said that anything that makes that much money couldn't be all legal. As they say, where there's lots of cash, there's lots of dirt along with it. Henry made it very clear; his contacts were all in his head. I guess this is what you would call a real head job.

The next morning, they were all up at seven, having a light breakfast of bagels and coffee. Jim fired up the engines, and they headed towards Miami. During the boat ride back, Jim explained to Henry that on his return to San Francisco, he'd prepare the necessary employment contract for his review and signature, after which he would start the wheels in motion. Ron was scheduled to take his California bar exam next week, so that challenge would officially be out of the way for them. They had already agreed that they would live on the yacht, and their biggest challenge would be finding a location for all their furnishings, so there are still a few little things to take care of before the move.

That morning Ron and Jim took the earliest flight out of Miami and arrived in San Francisco at eleven-thirty a.m. They walked in the door of their apartment just after noon, and Jim threw his luggage on the bed and went directly to the phone to call David at the office. He asked him how everything had gone. He told Jim that everything was great and that Lee was just getting ready to leave the office. Jim told David to transfer the call to Lee so he could speak with him.

Lee picked up the phone and greeted Jim. Jim asked him how things had gone in his absence. Lee said, just great; I didn't know that you had such an efficient assistant, no wonder your job is getting easier by the day. Jim indicated that David had matured into another outstanding attaché. Lee said we have a lot to talk about; my replacement has been named and will assume my position in two weeks. Let's meet at ten o'clock tomorrow morning, and we'll go over all the details.

Before they hung up, the call was transferred back to David in which Jim asked him, after he got off of work tonight, if he wanted to stop by his house for Jim had a few things he wanted to discuss with him.

After they hung up the phone with the office, Ron and Jim sat down and had a detailed discussion as to their plan of action for the upcoming weeks. They decided that Ron would concentrate 100% of his time on everything related to his law degree and taking the bar exam for the state of California. Jim would assume all of the responsibility for relocating to Florida, terminating their lease, and taking care of their personal belongings. Jim told Ron that David was coming over tonight to join them for dinner and that they should offer him a position to work with them as their personal assistant. Ron was delighted to hear that proposal because David was so intelligent and protective of both of them; he would love having David around. Ron did bring up the fact, though, that they were going to live on the yacht, and wouldn't it be kind of cramped quarters? Jim said if he remembered correctly, there was a crew quarter with bunk beds tucked away by the engine room. Not to worry, I'm sure we can make proper accommodations.

David showed up at their apartment a bit after six-thirty. Jim had ordered dinner to be delivered so that they could all sit together and discuss their futures. They went over all of the details of Henry's global operation with David and told him that Ron and Jim would need his assistance if we were going to take over operations. Jim asked if he would like to join them and move to Miami? David planted that gorgeous grin on his face and let out a small chuckle and said, "Do you think that I was ever going to leave you guys?"

As they were talking about the details of their move, the phone rang. It was Henry from Miami. Henry was so excited that Jim could barely make out who it was at first. Jim finally said, "Henry take a deep breath, I can hardly understand you." Well, in his next breath, he said that his neighbors two doors down were selling a ninety-foot yacht. Well, I just bought it. I knew "TITO" would not be big enough to be your floating office and home while going to the remote locations here in the Caribbean Sea to meet with the exporters in our operation.

Jim told Henry that he didn't know if they could afford that large a yacht. Henry says, "Don't worry about that; I'll take care of all that. I'm going to name it the "TITO II." It will be relocated to my dock this week, and I'll have it cleaned, detailed, and ready for you to move in within two weeks. Mohammed, the current owner of the yacht, described all the details to me. Well, it's kind of ostentatious, Henry said, I will have to give you a quick overview because there's just too much for me to describe:

1. Four master staterooms with full baths
2. Two crew quarters with adjoining bath, both have bunk beds
3. Captain's quarters with adjoining bath
4. Full galley with all Bosch appliances
5. Dining room that can seat sixteen
6. The main salon that seats sixteen
7. Four decks - one deck is decorated as a man cave
8. Powered by two Caterpillar engines with a cruise speed of 30 km
9. Comes with the current crew of four

"That's just the highlight, boys, do you think that will meet your specifications?"

"Henry," Jim said, "I don't know what to say, you're kinder to me than my own father had ever been. My team is here in the dining room tonight talking about our plans, and it looks like we should be there within two to three weeks; I'll be in touch."

After Jim got off the phone, the three of them were in a state of absolute shock. Now they knew that their decision to go to Miami and run the business was the right decision to make. After all this time, Tito's plans were finally materializing. Realizing that Tito did all this for him, Jim promised to himself that he was going to take care of the business and Tito's parents to the best of his ability.

After the phone call, they all got down to business. As discussed, Ron would take care of anything and everything that had to do with his law degree. Jim asked David if he would continue to take care of operations at the bank, while he took care of relocating their personal property and locating storage in Miami for their household furnishings for everything that they would not need on the yacht.

On Monday, Jim met with Lee at ten o'clock. Lee told Jim that his replacement was Larry Packard and would be reporting in two weeks. Then Jim gave Lee two letters of resignation, one From David and the other from Jim, effective in two weeks. Lee said, "Well, I guess since we're going to clean out the office, I better inform Larry of the decisions so that he can make arrangements for his own staff."

Lee and Jim went to lunch almost every day during those two weeks. Their goodbyes were sometimes tearful, but they were both pleased about their futures. They assured each other that they were just a phone call away and how much they had enjoyed working together over the last fifteen years.

On the very last day, Lee gave Jim a gold Rolex watch engraved with "You're the best." Jim was so moved that he gave Lee a tearful hug and thanked him from the bottom of his heart for everything that he had done for him over the past fifteen years.

MIAMI

The three of them flew into Miami and arrived during a blistering hot Monday morning on July 1, 1985. Henry's driver met them at the airport and took them directly to his estate. He and Mary were standing at the door, waiting to greet them with open arms. Jim and Ron introduced them to David, and they welcomed him in as they did them in the past. They then proceeded walking them through the house down the walkway to the pier and to see what was the most outstanding yacht you could ever lay your eyes on. Standing alongside, in their white attire, were the Captain and his crew.

1. Captain John
2. Chef Timothy
3. Stuart Andrew
4. Deckhand Stephen

After Jim, Ron, and David introduced themselves, the staff immediately took their personal belongings below deck, and Captain John gave them a private tour of the ninety-foot yacht. After all the formalities, the five of them retired to the back deck for cocktails. Henry told them that he had transferred all the electronic information from his old yacht to the new one. He had also set up an office for them on the top deck with a panoramic view and a 360° radius, letting them see if they were being approached by other boats at all angles.

They continued sipping on cocktails until dinner hour. Promptly at seven, Andrew summoned everyone to the dining room, and Chef Timothy prepared a dinner of filet mignon and lobster with grilled asparagus and a beautiful arugula salad. It was so delicious that the three of them from San Francisco were in complete awe. Jim said to Ron and David, "If this is what we get to look forward to over the next coming months were going to gain a lot of weight." There was a chuckle around the table.

After dinner, the group all adjourned to the salon, where the air conditioning had it cooled to a comfortable seventy-four degrees. They continued pleasantries and reminisced on the old times. Henry and Mary excused themselves and retired to their private quarters at their estate while Ron, David and Jim went up to the top deck and relaxed under the stars. They summoned Stephens' to bring them a drink and told him he was excused for the remainder of the evening. Then the three of them started to thank their lucky stars for where they were today and were looking forward to their now bright futures together.

EL COMERCIO

The next morning at nine o'clock, they all went to the office on the top deck. This deck was massive; it had four desks, a computer center, a fax/copy machine, and four telephone lines plus a video screen with satellite cable television. It also had a fully stocked wet bar and refrigerator. The Captain's pilothouse was directly ahead but was connected by way of an enclosed circular stairway from the deck below, in which he gained entrance to his office, so he did not have access to Jim and Ron's office, which was off-limits to the crew.

Jim first asked Ron if he planned on taking the Florida bar exam since they are now going to be staying here in Miami. He told them that he had passed the California bar exam and was scheduled to take the Florida bar exam in two weeks. He had hired a tutor to prepare for the exam, and he will be coming to the yacht daily for private tutoring sessions.

David and Jim got down to business. The television that Henry had installed from the previous yacht pinpointed all of their offices around the Caribbean Sea. David was taken back by the massive operation that they had just undertaken and said that for many of the locations, he and Jim should have some working knowledge of already because they had banks there. Jim and Ron agreed and thought that would make their jobs a lot easier.

They discussed the need to visit as many of the locations in the Caribbean Sea as soon as possible. A list of the ports included:

1. Port Angelo Kingston in Jamaica
2. Port Colon in Panama
3. Port Santo Domingo in Dominican Republic
4. Port of La Guaira in Venezuela
5. Port Geopec Toulon in the Grand Caymans
6. Puerto Barrios Guatemala

Jim said to Ron and David, "We will stay in port until Ron completes his bar exam next week, then we will set sail." Jim asked David to prepare a memorandum and to wire it to all of the port managers, which would introduce them as their new management staff. They broke for lunch and went to the second deck, which was like a man cave. It had a wet bar, a game table/lunch table, a thirty-six-inch television, and a leopard skin rug. Timothy whipped up another delicious lunch, and afterward, they all sat back, enjoying the air-conditioning of the man cave while sipping on a cool drink. Jim, Ron and David all decided to call it a day; they did not want to overdo it on their first day on the job.

They continued to talk and had a few more drinks when David spoke up. "So… When are you and Ron going to get married?" Jim replied, "Well, David, you have always told me that you wanted to go to the seminary, correct. Now personally, I think that would be such a waste of your skills, for you are already a divine spirit. But when you have the authority to marry someone, we would like to be your first ceremony. And we would like to have it on board the "TITO II."

Ron immediately spoke up, "Jim, you did not even ask me to marry you…" Jim, with a puzzled expression on his face, replied, "Oh, I completely forgot that little detail Ron, "Will you marry me"?"

Ron hopped up emphatically and replied, "I thought you would never ask; of course, I will, yes!"

David spoke up once more, "Jim, I have to honestly tell you this because the three of us are going to be working so closely together, and I cannot keep this secret any longer. I have loved you, Jim, since the very first day I set my eyes upon you. Not in a sexual way but in a platonic loving way. As time passed, I have put my arms around you and surrounded you with all of the knowledge and protection I could muster. I love your relationship with Ron. I am in no way jealous; I'm just the opposite, extremely happy. I will do everything within my power to contribute to your happiness together and ensure our operation is a success so we can then live our lives as we deserve." Jim replied, "Ron and I feel the same way and are happy that we are family." Ron responded, "No, I don't. Get Out! I'm just joking with you, get over here," and as they all embraced each other, Ron told Jim and David that he loved them both.

BLUE WATERS

$\mathcal{I}$im and his crew pulled out of Miami on the first day of September on a bright sunny day. Their first stop was to the Grand Caymans. There, as with the rest of the ports in the Caribbean Sea, they met with the port managers and staff. Ron conducted a legal review of all of our clients' contracts as well as made several copies. David reviewed all of the computerized information on file and downloaded it on floppy disks that they took with them. Jim both reviewed and took the copies of the financial information and supporting documents that Ron provided him. They would never meet any of the management teams on their yacht. All business was conducted at their office at the ports. When their jobs were completed, they motored onto the next port of call.

Captain John and the crew would routinely have the yacht serviced and refitted while Jim, Ron and David were in each port. While they were off on business, Jim granted the staff time off, of course, only after all of their duties were completed in full. Jim was at sea for approximately ninety-five days. During this time, the yacht performed beautifully, and the interior looked as nice as it did the day they left Miami. After the yacht was thoroughly cleaned and refitted, the crew had two weeks off with pay, before returning to work just before Christmas eve.

Their first Christmas on the yacht was a wonder to behold. Henry, Mary, Ron, Dave, Jim, and the ship's crew celebrated in a big way. The yacht was outfitted with Christmas lights and a Christmas tree befitting the yacht's extravagant interior. On Christmas Eve, they partied until the stroke of midnight, and at that time, gifts were exchanged. Ron and Jim gave their crew new specially designed uniforms with the insignia of the Tito II emblazed in embroidery with gold lettering and a $2,000 check. They also gave David a gold Rolex with diamonds with the inscription "To the one we love," along with a check for $5,000. Ron and Jim's gift to each other was their undying love and understanding. Henry and Mary gave Ron and Jim the title to the yacht, and Ron immediately started the paperwork to have the "TITO II" registered as an LLC.

On Christmas day, Jim summoned the caterer to prepare a Christmas dinner for everyone, and it was enjoyed to the fullest. After the crew was dismissed, they spent the remainder of the afternoon on the yacht giving thanks, many times to Tito for the blessing he had bestowed on all of them and for his wish of Jim taking over his father's import/export business.

During the first week after they returned, Ron, David and Jim spent time consolidating and reviewing all of the information that they had collected from their units in the Caribbean Sea. They were shocked to learn that the combined net profit after all expenses, including theirs, for the global operation, totaled over $10,000,000 a month in net profit. The only money that was visibly transferred to Henry &/or Mary Martinez, was in the form of a check. It was the dock rent collected to the sum of $4,000,000 a month. (Or was it retirement pay?)

The following units contributed to the profits

1. Jamaica — $1.0 Million
2. Panama — $2.0 Million
3. Dominican Republic — $1.5 Million
4. Venezuela — $3.0 Million

5.	Grand Cayman	$500,000
6.	Guatemala	$2.0 Million
TOTAL		$10,000,000

Astounded as to the profitability of the operation, Ron and Jim knew perfectly well that this type of profit could not be made entirely from a legal operation. It has now been abundantly clear that their new challenge was to make sure that all the documentation on file did not connect them or Henry to any of the perceived illegal operations.

During that week, they started to research in detail each of the profit centers. They noticed that in Costa Rica that they had two of the company's planes in the air all the time flying in and out of Peru. The same went for Columbia; planes were transporting cargo to Guatemala daily. The cargo ships out of these ports were heading to Europe. Ships out of the Dominican Republic went to Miami and up the coast to New York. They had a total of eight large planes and ten cargo ships on the move at all times.

As New Year's Eve was rapidly approaching, Captain John asked if they had developed any plans to celebrate their first New Year's on board the yacht.

Jim replied that they had made no specific plans and asked the Captain what he had in mind. The Captain said that the four of them all had a special friend that they would like to invite to join them for a New Year's Eve celebration if they all didn't mind. Ron, David and Jim got together and concluded that it would be fun to have some strangers on board and meet the friends of their crew.

John, Timothy, Andrew, and Stephen all had boyfriends onshore. We were surprised when they arrived on board about six o'clock on New Year's Eve. Timothy had contacted the caterers to provide a full-blown spread for them to enjoy throughout the evening. All of their new companions were extremely handsome and exciting to be with. It was strange to see the crew affectionately embracing their loved ones in their boss's presence. Jim, Ron and David welcomed the New Year's with a real bang, and after which, the crew retired with their guests to their private quarters. After everyone had gone, the three of them continued their evening sitting under the stars. For the very first time since Ron and Jim had known David, they invited him to join them for the night. David graciously accepted their invitation. As Ron and Jim's relationship was monogamous, which

David understood, the three of them all crawled into bed and cuddled until they fell to sleep.

The next morning, they had breakfast, and the guests departed the yacht while the crew went back to work performing their respective duties. David expressed his appreciation for Jim and Ron's love and understanding and wanted to tell them how much he loved them. "Well, it's a new year, we have much to do, places to go, and information to analyze." "And many questions to be answered about the legality of our business empire," Jim thought to himself.

THE STORY

*M*any of you may ask, how is this a Vietnam Love Story? Well, to begin:

1. Before 1968 three strangers set the stage, not knowing that their paths would later converge.

2. By the time of 1968, the U.S. invested three years in the war in Vietnam.

3. By chance, Tito, Jim and Ron were all enlisted into the Army in 1968.

4. All three of their paths converged in similar fields and close proximity in Vietnam.

5. Tito and Jim fell in love with each other at first sight that year.

6. Jim ended up meeting Ron and becoming very close friends with him almost instantly.

7. More than a decade later, after the war, Vietnam's pull never left Jim, for he and Ron ended up falling in love.

This all adds up to an interweaving story of love during a period of intense war in Vietnam. Within spending eight months of their life together, what Tito and Jim shared between each other was more important than owning a degree in finance from Harvard or even a lifetime's worth of love therapy. In the most un-Orthodox environment, surrounded by war, each found love beyond their wildest expectations. After a little over a decade, Ron confessed his love he secretly held for Jim all those years after they first met, and from there on, their romance blossomed on.

IN THE NEXT BOOK

Onboard the "Tito II," Jim, Ron and David launched a full investigation of the business, making sure every detail was examined. Any possible underworld involvement would be exposed. And the details of that prolonged investigation would be revealed in full to the Martinez's.

At the same time, Jim and Ron exchanged their vows onboard the "Tito II," which was conducted by Captain John. David was their best man; Jared and Jenkins flew in from Tulsa, Oklahoma, to attend their ceremony.

As their happiness grows, so does the revelation of dark secrets, hidden dangers, and a web of felonious activities threatening to engulf them all.

To be continued…

FORGOTTEN LOVE STORIES

*B*etween 1964 and 1975, 47,434 American soldiers lost their lives during the war in Vietnam. Even if only half of a percent of those casualties experienced a similar type of forbidden love while serving our nation, that means there are still at least 237 untold tales of love lost. Only one of those stories was told today, though fictionalized, the story was influenced by some events that took place before, during, and after the author's tour in Vietnam. If you or a friend, too, had experienced this type of love or loss and you wish to share your experience, please contact the author, James Marquis, at jamesmarquis68@gmail.com; he will fictionalize your story to celebrate your love.

ABOUT THE BOOK

"1968 A Vietnam Love Story" is a fictional story of the forbidden love of several soldiers during the Vietnam War. Their story of finding love and happiness spans over two decades before, during, and after the Vietnam War.

The main characters are tragically reminded that their loved one's safety is never guaranteed no matter where they are located. During war, tragedy and heartbreak can lead to never-ending love and ultimate happiness.

A whirlwind of global intrigue surrounded by a monogamous love affair delves into the illegal world of high finance. The author shares with you the innermost feelings of each character, and hopefully, you will enjoy the journey.

James Marquis

ABOUT THE AUTHOR
James Marquis

James was born the son of sharecroppers and grew up in tenement housing on a farm in Iroquois County, Illinois. At the age of thirteen, he and his family moved to a small village in the same county where he graduated high school. At the age of seventeen, he moved from the small village to Champaign, Illinois, the home of the University of Illinois, and he secured a position as a teller at one of the major banks. Because of his "Personality, Looks and Communication skills," it was not long before he was promoted to head teller. It was during this time frame he was drafted into the Army to supplement the surge of troops during the Vietnamese crisis.

His distinguished military career was recognized by being awarded a Bronze Star Medal. His training in the Army prepared him to pursue his life's dream of acquiring financial freedom. After being discharged from the Army, he returned to the small Illinois bank and discovered it had limited promotional opportunities for him to achieve his dream.

James sought out and attained a position with one of the world's largest banks based in California, where he spent 25 years. During this time, he was recognized internationally as the bank's top motivational speaker. During the last five years of his employment, he was attaché to the chief financial officer of international banking.

He was influenced to put his writing talents in book form by the author named Annie Prouix, who wrote the book "Brokeback Mountain." James has also written a fictional story entitled "1862: A Civil War Love Story".

DISCLAIMER TO THE READER
"1968 A Vietnam War Love Story"